Texts in Computer Science

Series Editors

David Gries, Department of Computer Science, Cornell University, Ithaca, NY, USA

Orit Hazzan, Faculty of Education in Technology and Science, Technion—Israel Institute of Technology, Haifa, Israel

Sergei Kurgalin • Sergei Borzunov

Concise Guide to Quantum Computing

Algorithms, Exercises, and Implementations

 Springer

Sergei Kurgalin
Digital Technologies
Voronezh State University
Voronezh, Russia

Sergei Borzunov
Digital Technologies
Voronezh State University
Voronezh, Russia

ISSN 1868-0941   ISSN 1868-095X   (electronic)
Texts in Computer Science
ISBN 978-3-030-65054-4   ISBN 978-3-030-65052-0   (eBook)
https://doi.org/10.1007/978-3-030-65052-0

This Springer imprint is published by the registered company Springer Nature Switzerland AG
The registered company address is: Gewerbestrasse 11, 6330 Cham, Switzerland

Preface

The role of Computing Science in modern society can hardly be overestimated. The rapid development of physics, chemistry, biology, medicine as well as achievements in economics is based on the developing possibility of solving resource-intensive problems on powerful computing systems. Nevertheless, the exponential complexity of many practically significant problems leads to considerable difficulties even while using a supercomputer.

The approach to computing, fundamentally different from the widely known and available—classical—computing, provides quantum theory. Quantum-mechanical methods of solving problems in some cases provide exponential speedup in comparison with traditional algorithms. The idea of accelerating computational processes by using quantum systems began to develop in many academic and industrial world centers.

Training specialists in the field of computing science, meeting the requirements of time, leads to the need to pay special attention to the basics of quantum computing. In our opinion, there are not enough books describing quantum computing on the methodical level available for future programmers in the educational literature at the present time. The known textbooks on quantum theory in the majority do not contain the data on quantum algorithms and, therefore, are not completely focused on the training of experts in the field of information technology. In this workbook, we made an attempt to fill this gap.

The book was developed over several years using the experience of teaching the quantum theory course at Computer Science Faculty of Voronezh State University. Its content corresponds to Federal State Educational Standards in the areas of Bachelor's Degree "Information Systems and Technologies", "Software Engineering", "Information Security", "Mathematics and Computer Science", Specialist's degree in "Computer Security", and Master's degree in "Information Systems and Technologies".

This workbook is intended for laboratory, practical classes and for independent study. It contains basic theoretical concepts and methods for solving basic types of problems, as well as a large number of explanatory examples. It gives an idea of the quantum computing model, basic operations with qubits and universal elements of quantum schemes, introduces definitions of entangled states, and analyzes the Einstein–Podolsky–Rosen experiment. Much attention is paid to examples of the most important quantum algorithms—the algorithms of phase evaluation and the Fourier quantum transformation.

The workbook offers a large number of problems and exercises of a wide range of complexity for independent solution. All problems, except the simplest ones, are equipped with answers, hints, or solutions. The most difficult problems are highlighted with the character "asterisk" (*), placed before their number. The end of an example in the text is indicated by the symbol □. The book includes a sufficient number of illustrations and diagrams to help visualize the objects being studied and the connections between them. Review questions on theoretical material are formulated to test knowledge.

There are two appendices to the textbook. In *Appendix A*, the system of postulates of quantum theory is formulated. That will help students to fix and systematize knowledge and skills in solving quantum-mechanical problems. *Appendix B* contains a summary of information on Hermitian and unitary transformations, which are known to play a major role in quantum theory and its applications.

Mastering the material requires knowledge of the basics of linear algebra and elementary probability theory. The knowledge of quantum mechanics basics is not strictly necessary.

The authors sought to make the presentation of the material available without losing the rigor of the wording of the statements. Theorems and properties are accompanied by proofs or references to scientific literature.

A detailed reference and bibliography aids is one of the distinguishing features of this manual. *References* contains links to educational literature, where the issues discussed in the text of this workbook are covered in detail. In addition, the reference list contains references to fundamental original works in the field of quantum computing. It is hoped that the reading of such articles will allow students to get the skill of working with a steadily increasing flow of information about modern researches in this field.

The workbook is provided with *Name* and *Subject Indices*.

Below you can see the scheme of the chapter information dependence in the form of an oriented graph reflecting the preferable order of covering the academic material. For instance, after having studied Chaps. 1–3, you can move to one of the three Chaps. 4, 5 or 6, whose contents are relatively independent. The dashed border marks Chap. 7, which will be suitable for readers who want to reach a high level of the subject knowledge; this section contains more difficult academic material. This way, after studying Chap. 6, you can either come to Chap. 8, or, for better mastery, study the material of Chap. 7 more thoroughly for better digestion, and only then switch to Chap. 8.

Voronezh, Russia   Sergei Kurgalin
July 2020   Sergei Borzunov

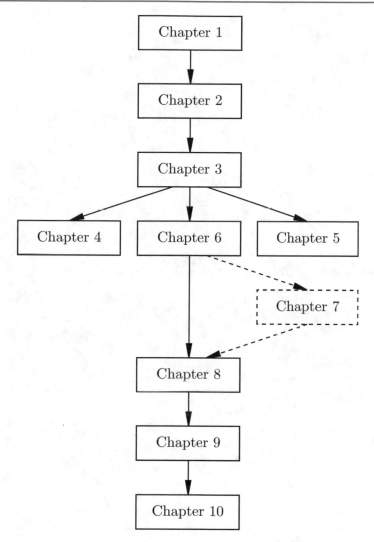

The chapter dependency chart

Acknowledgments

The authors would like to express their grateful acknowledgment to their colleagues Valeria Kaverina, Alexander Klinskikh, Alexey Kornev, Sergei Kurgansky, Peter Meleshenko, Mikhail Semenov, and Sergei Zapryagaev for useful discussions, advice, and critical comments.

We would like to thank Senior Editor Wayne Wheeler from Springer. Editorial Director Aliaksandr Birukou provided continuous and comprehensive support at all stages.

Probably, there are inevitable errors and inaccuracies in the book, all of which remain on the authors' conscience.

Sergei Kurgalin expresses special gratitude to Olga Kurgalina, Alexander Shirokov, and Veronica Shirokova for their constant support during the work on the book.

July 2020
Sergei Kurgalin
Sergei Borzunov

Contents

Notation

$\lvert\psi\rangle$	Quantum state, ket vector
$\langle\psi\rvert$	Bra vector
$\lvert 0\rangle, \lvert 1\rangle$	Basis quantum states of the qubit
$\lvert\psi_1\rangle \otimes \lvert\psi_2\rangle$ or	Tensor product
$\lvert\psi_1\rangle\lvert\psi_2\rangle$	
$A = (a_{ij})$	Matrix, formed by elements a_{ij}
A^T	Transposed matrix
$A^\dagger$	Hermitian conjugate matrix
M_U	Matrix of a unitary transformation U
δ_{ij}	Kronecker delta
$\sigma_1, \sigma_2, \sigma_3$	Pauli matrices
$\alpha_1, \alpha_2, \alpha_3, \beta$	Dirac matrices
$\lambda_1, \lambda_2, \ldots, \lambda_8$	Gell-Mann matrices
$\{a_1, a_2, \ldots, a_n\}$	The set consisting of the elements $a_1, a_2, \ldots, a_n$
$\mathbb{N} = \{1, 2, 3, \ldots\}$	The set of natural numbers
$\mathbb{Z} = \{0, \pm 1, \pm 2, \ldots\}$	The set of integers
$\mathbb{R} = (-\infty, \infty)$	The set of real numbers
$\mathbb{C} = \mathbb{R} \times \mathbb{R}$	The set of complex numbers
$\mathbb{B} = \{0, 1\}$	Two-element set
$\mathbb{H}$	Hilbert space
$A \Rightarrow B$	The logical implication
$A \Leftrightarrow B$	The logical equivalence
$\forall x(P(x))$	For all x, the statement $P(x)$ is true
$\exists x(P(x))$	There exists such x, that the statement $P(x)$ is true
$[a, b]$	Closed interval $\{x : a \leqslant x \leqslant b\}$
(a, b)	Open interval $\{x : a < x < b\}$
$i = \sqrt{-1}$	The imaginary unit
z^*	Conjugate of a complex number z
$\sum_{i=1}^{n} a_i$	Sum $a_1 + a_2 + \ldots + a_n$
$\prod_{i=1}^{n} a_i$	Product $a_1 a_2 \ldots a_n$
I	Identity matrix
O	Zero matrix
$[A, B]$	Commutator of matrices A and B

$\{A, B\}$	Anticommutator of matrices A and B
$(a_1, a_2, \ldots, a_n)$	Vector, i.e. an element of the set A^n
$\lfloor x \rfloor$	Floor function of x, that is the greatest integer less or equal to the real x
$x_1 \wedge x_2$	The conjunction of Boolean values x_1 and x_2
$x_1 \vee x_2$	The disjunction of Boolean values x_1 and x_2
$x_1 \rightarrow x_2$	The implication, x_1 implies x_2
$x_1 \leftrightarrow x_2$	The equivalence of x_1 and x_2
$x_1 \oplus x_2$	Addition modulo 2
$x_1 \mid x_2$	Sheffer stroke
$x_1 \downarrow x_2$	Peirce's arrow
$O(g(n))$	The class of functions growing not faster than $g(n)$
$\Omega(g(n))$	The class of functions growing at least as fast as $g(n)$
$\Theta(g(n))$	The class of functions growing of the same order as $g(n)$

Notation of main quantum gates see on pages 13 and 30

Quantum Computing Model

1

Quantum computers use processes of a quantum nature manifested with atoms, molecules, molecular clusters, etc. The description of such processes is based on the application of complex numbers and complex matrices.

As is well known, the basic notion of classical information theory is a **bit** [1]. A classical bit takes the values 0 or 1 (and no other).

A qubit (quantum bit) is the smallest element that executes the information storage function in a quantum computer [2].

Qubit is a quantum system $|\psi\rangle$ that allows two states: $|0\rangle$ and $|1\rangle$. In accordance with the so-called "bra-ket" Dirac[1] notation (from word *bra(c)ket*), the symbols $|0\rangle$ and $|1\rangle$ are read as "Ket 0" and "Ket 1", respectively. The brackets $|\dots\rangle$ show that ψ is some state of the quantum system.

The fundamental difference between the classical bit and the qubit consists in that, the qubit can be in a state different from $|0\rangle$ or $|1\rangle$. The arbitrary state of the qubit is defined by the linear combination of basis states:

$$|\psi\rangle = u\,|0\rangle + v\,|1\rangle,\tag{1.1}$$

where the complex coefficients u and v satisfy the following **normalization condition**:

$$|u|^2 + |v|^2 = 1.\tag{1.2}$$

The mathematical description of basis states reduces to their representation in matrix form:

$$|0\rangle = \begin{bmatrix} 1 \\ 0 \end{bmatrix}, \quad |1\rangle = \begin{bmatrix} 0 \\ 1 \end{bmatrix}.\tag{1.3}$$

Based on the presentation (1.3), the arbitrary state of the qubit is written as

$$|\psi\rangle = \begin{bmatrix} u \\ v \end{bmatrix}.\tag{1.4}$$

---

[1] Paul Adrien Maurice Dirac (1902–1984).

© The Author(s), under exclusive license to Springer Nature Switzerland AG 2021
S. Kurgalin and S. Borzunov, *Concise Guide to Quantum Computing*,
Texts in Computer Science, https://doi.org/10.1007/978-3-030-65052-0_1

A system of two qubits is set by a linear combination of basis states

$$|00\rangle = \begin{bmatrix} 1 \\ 0 \\ 0 \\ 0 \end{bmatrix}, \quad |01\rangle = \begin{bmatrix} 0 \\ 1 \\ 0 \\ 0 \end{bmatrix}, \quad |10\rangle = \begin{bmatrix} 0 \\ 0 \\ 1 \\ 0 \end{bmatrix}, \quad |11\rangle = \begin{bmatrix} 0 \\ 0 \\ 0 \\ 1 \end{bmatrix}. \tag{1.5}$$

Similarly are introduced the states

$$|00\ldots00\rangle, \quad |00\ldots01\rangle, \quad \ldots, \quad |11\ldots11\rangle \tag{1.6}$$

of several interacting qubits. Such quantum states are called **computational basis states** or, for short, **basis states**.

In general, the state of a quantum system is described by a vector $|\psi\rangle$ of the **Hilbert**[2] **space** $\mathbb{H}$. The Hilbert space is a natural generalization of a concept of finite-dimensional vector space $\mathbb{C}^n$ to infinitely large n [3–5].

Example 1.1 Consider two qubits brought to the states $|\psi_1\rangle = u_1 |0\rangle + v_1 |1\rangle$ and $|\psi_2\rangle = u_2 |0\rangle + v_2 |1\rangle$, respectively. Let us determine the state of the system formed by these qubits as a whole.

Solution.

If two qubits are in the states $|\psi_1\rangle$ and $|\psi_2\rangle$, respectively, the state of this quantum system as a whole is described by the product of vectors, or, in other words, of **wave functions**:

$$|\psi_1\rangle |\psi_2\rangle = (u_1 |0\rangle + v_1 |1\rangle)(u_2 |0\rangle + v_2 |1\rangle)$$
$$= u_1 v_1 |00\rangle + u_1 v_2 |01\rangle + u_2 v_1 |10\rangle + v_1 v_2 |11\rangle. \tag{1.7}$$

$\square$

Note. More strictly, the formula (1.7) refers to the so-called **tensor product** in the Hilbert space. For the tensor product, it is often used as the designation "$\otimes$": $|\psi_1\rangle \otimes |\psi_2\rangle$. The state of the composite system is given by the tensor product of its components. For instance, if a system is formed by three isolated subsystems in states ψ_1, ψ_2, and ψ_3, then the composite system according to quantum mechanics principles is described by the vector $|\psi_1\rangle \otimes |\psi_2\rangle \otimes |\psi_3\rangle$, or, for short, $|\psi_1\rangle |\psi_2\rangle |\psi_3\rangle$.

The designation $\langle\psi|$—"bra" vector—applies to the Hermitian conjugate of the vector $|\psi\rangle$ [6, 7]:

$$\langle\psi| = (|\psi\rangle)^{\dagger}, \tag{1.8}$$

and vice versa:

$$|\psi\rangle = (\langle\psi|)^{\dagger}. \tag{1.9}$$

Let us remember that for an arbitrary complex matrix Z of size $m \times n$, containing elements $Z = (z_{ij})$, where $i = 1, 2, \ldots, m$, $j = 1, 2, \ldots, n$, the elements of the Hermitian conjugate matrix are equal to

$$z_{ij}^{\dagger} = z_{ji}^{*}, \tag{1.10}$$

---

[2]David Hilbert (1862–1943).

where z^* is a complex conjugate of z (see Appendix B on page 103).

The vector, dual to $c\,|\psi\rangle$, is $c^*\,\langle\psi|$.

For computational basis vectors, we have the following representations in the matrix form: $\langle 0| = [1\ 0]$, $\langle 1| = [0\ 1]$.

It is also easy to write down an arbitrary qubit $|\psi\rangle = u\,|0\rangle + v\,|1\rangle$:

$$\langle\psi| = \begin{bmatrix} u \\ v \end{bmatrix}^{\dagger} = [u^*\ v^*] = u^*\,\langle 0| + v^*\,\langle 1|. \tag{1.11}$$

The operation of the **scalar product** of complex vectors $|\psi_1\rangle = u_1\,|0\rangle + v_1\,|1\rangle$ and $|\psi_2\rangle = u_2\,|0\rangle + v_2\,|1\rangle$ is introduced according to the formula

$$\langle\psi_1|\psi_2\rangle = [u_1^*\ v_1^*] \times \begin{bmatrix} u_2 \\ v_2 \end{bmatrix} = u_1^* u_2 + v_1^* v_2. \tag{1.12}$$

Note that the formula (1.12) is a product of a row and a column well known from linear algebra. The scalar product is also known as the **dot product**.

Let us list the main properties of the scalar product. For arbitrary vectors $|a\rangle$, $|a_1\rangle$, $|a_2\rangle$, $|b\rangle$ and an arbitrary complex constant $\alpha \in \mathbb{C}$, the following relations are valid:

(1) $\langle a|a\rangle \geqslant 0$, where $\langle a|a\rangle = 0 \Leftrightarrow |a\rangle = 0$ (non-negativity);
(2) $\langle a|b\rangle = \langle b|a\rangle^*$ (conjugate symmetry);
(3) $\langle a_1 + a_2|b\rangle = \langle a_1|b\rangle + \langle a_2|b\rangle$ (anti-linearity with respect to the first argument);
(4) $\langle\alpha a|b\rangle = \alpha^*\langle a|b\rangle$ (anti-linearity with respect to the first argument).

The system of vectors $\{|e_1\rangle, |e_2\rangle, \ldots, |e_n\rangle\}$ is called **orthonormal**, if the condition

$$\forall i, j = 1, 2, \ldots, n : \quad \langle e_i|e_j\rangle = \delta_{ij} \tag{1.13}$$

is met. Here, the designation δ_{ij} is entered for the **Kronecker**[3] **symbol**, defined as follows:

$$\delta_{ij} = \begin{cases} 1, & \text{if } i = j, \\ 0, & \text{if } i \neq j. \end{cases} \tag{1.14}$$

In other words, the Kronecker symbol is equal to one in case of matching indexes, and zero if the indexes are different.

For two arbitrary states of n-qubit systems

$$|\psi'\rangle = \begin{bmatrix} u_1 \\ u_2 \\ \vdots \\ u_{2^n} \end{bmatrix} \quad \text{and} \quad |\psi''\rangle = \begin{bmatrix} v_1 \\ v_2 \\ \vdots \\ v_{2^n} \end{bmatrix}, \tag{1.15}$$

the normalization condition is as follows:

$$\langle\psi'|\psi'\rangle = \sum_{i=1}^{2^n} |u_i|^2 = 1, \quad \langle\psi''|\psi''\rangle = \sum_{i=1}^{2^n} |v_i|^2 = 1. \tag{1.16}$$

---

[3] Leopold Kronecker (1823–1891).

The orthogonality condition of two states $|\psi'\rangle$ and $|\psi''\rangle$ is written as follows:

$$\langle\psi'|\psi''\rangle = \sum_{i=1}^{2^n} u_i^* v_i = 0. \tag{1.17}$$

Note that the states of the computational base (1.6) are orthonormalized.

To change the state of a quantum system, quantum operations are used, which are called **quantum logic gates**, or, for short, simply **gates**. Thus, gates perform logical operations on qubits. Note that the change of state $|\psi\rangle$ in time is also referred to as the **evolution** of the quantum system.

An important step of quantum algorithms is the procedure of **measurement** of a state. When the qubit state is measured, it randomly passes to one of its states: $|0\rangle$ or $|1\rangle$. Therefore, the complex coefficients u and v from the qubit definition (1.1) are associated with the probability to get the value 0 or 1 when its state is measured. According to the postulates of quantum theory, the probabilities of passing to the states $|0\rangle$ and $|1\rangle$ are equal to $|u|^2$ and $|v|^2$, respectively. In this connection, the equality (1.2) reflects the probability conservation law. After the measurement, the qubit passes to the basis state, complying with the classical result of the measurement. Generally speaking, the probabilities of getting the result 0 and 1 are different for different states of the quantum system.

In other words, quantum computing is a sequence of simple form operations with the collection of the interacting qubits. In the final step of the quantum computing procedure, the state of the quantum system is measured and a conclusion about the computing result is made. The measurement makes it possible to obtain, at a macroscopic level, the information about the quantum state. The peculiarity of the quantum measurements is their irreversibility, which radically differentiates quantum computing from the classical one.

Despite the fact that the number of qubit states is infinite, with the help of measurement it is possible to obtain only one bit of classical information. The measurement procedure transfers the qubit state to one of the basis states, so a second measurement will produce the same result.

Quantum computer is a set of n qubits controlled by external (classic) signals [8–12]. Often, an ordered set of some qubits is called a **register**. The main elements of a quantum computer are shown in Fig. 1.1.

The classical quantum computer setting consists of the controlling classical computer and impulse generators controlling the qubit evolution, as well as measurement instruments of the qubit state. The system from n qubits in the initial state, e.g. $|\psi_{in}\rangle = |00\ldots0\rangle$, forms a memory register prepared to record input data and perform computations. The data are recorded by an external action on each of the system's qubits. The solution of the problem is determined by a measurement of the final state qubits $|\psi_{out}\rangle$ [2, 12].

The state of the form $|b_{n-1}\ldots b_1 b_0\rangle$ is usually written in decimal notation as $|a\rangle$, where the number

$$a = 2^0 b_0 + 2^1 b_1 + \cdots + 2^{n-1} b_{n-1}. \tag{1.18}$$

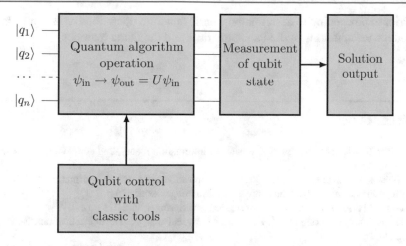

In this record, the state $|\psi\rangle$ of the n-qubit register of a quantum computer is expressed through the superposition of vectors of computational basis $B = \{|0\rangle, |1\rangle, \ldots, |2^n - 1\rangle\}$ by the formula

$$|\psi\rangle = \sum_{k=0}^{2^n-1} c_k |k\rangle, \tag{1.19}$$

where the normalization condition

$$\sum_{k=0}^{2^k-1} |c_k|^2 = 1 \tag{1.20}$$

is met.

A quantum system, formed by N two-level elements, has $\Sigma(N) = 2^N$ independent states. The key point of the functioning of such a system is the interaction of separate qubits with each other. The number of states $\Sigma(N)$ grows exponentially with the growth of the quantum system, which allows solving practical problems of a very high asymptotic complexity. For example, an efficient quantum algorithm of prime factorization is known, which is very important for cryptography [13]. As a result, the quantum algorithms provide exponential or polynomial speedup in comparison with the classical solution methods for many problems.

Unfortunately, no full-function quantum computer has been created yet, although many of its elements have already been built and studied at the world's leading laboratories [11, 14, 15]. The main obstacle to the development of quantum computing is the instability of a system of many qubits. The more the qubits are united into an entangled system, the more the effort is required to ensure the smallness of the number of measurement errors. Nevertheless, the history of quantum computer development demonstrates an enormous potential laid in the uniting of quantum theory and algorithm theory.

Prior to proceeding to describing the basic quantum operations with qubits, let us introduce the notations of the Pauli[4] matrices and the Dirac matrices.

References

1. Cover TM, Thomas JA (2006) Elements of information theory, 2nd edn. Wiley-Interscience, xxiii, 748 p
2. Nielsen MA, Chuang IL (2010) Quantum computation and quantum information. 10th anniversary edition. Cambridge University Press, Cambridge, xxxi, 676 p
3. Day MM (1973) Normed linear spaces, 3rd edn. Springer, viii, 211 p
4. Kolmogorov AN, Fomin SV (1975) Introductory real analysis. Dover Publications, New York, xii, 402 p
5. von Neumann J (2018) Mathematical foundations of quantum mechanics. Princeton University Press, 304 p
6. Dirac PAM (1939) A new notation for quantum mechanics. Math Proc Camb Philos Soc 35(3):416–418
7. Dirac PAM (1967) The principles of quantum mechanics, 4th edn. (Series: The international series of monographs on physics, 27). Oxford University Press, 314 p
8. Bennett CH, DiVincenzo DP (2000) Quantum information and computation. Nature 404:247–255
9. Galindo A, Martín-Delgado MA (2002) Information and computation: classical and quantum aspects. Rev Modern Phys 74(2):347–423
10. Harrow AW, Montanaro A (2017) Quantum computational supremacy. Nature 549:203–209
11. Ladd TD, Jelezko F, Laflamme R, Nakamura Y, Monroe C, O'Brien J (2010) Quantum computers. Nature 464:45–53
12. Valiev KA (2005) Quantum computers and quantum computations. Physics - Uspekhi 48(1):1–36
13. Shor PW (1997) Polynomial-time algorithms for prime factorization and discrete logarithms on a quantum computer. SIAM J Comput 26(5):1484–1509
14. DiVincenzo DP (2000) The physical implementation of quantum computation. Fortschritte der Physik 48:771–783
15. Williams CP (2011) Explorations in quantum computing, 2nd edn. (Series: Texts in computer science). Springer, xxii, 717 p

---

[4]Wolfgang Ernst Pauli (1900–1958).

Pauli Matrices and Dirac Matrices

2

For the imaginary unit $\sqrt{-1}$, we will use the designation $i \equiv \sqrt{-1}$. The matrices σ_1, σ_2, and σ_3:

$$\sigma_1 = \begin{bmatrix} 0 & 1 \\ 1 & 0 \end{bmatrix}, \quad \sigma_2 = \begin{bmatrix} 0 & -i \\ i & 0 \end{bmatrix}, \quad \sigma_3 = \begin{bmatrix} 1 & 0 \\ 0 & -1 \end{bmatrix}, \tag{2.1}$$

are called the **Pauli matrices**. They are widely used in quantum theory for describing half-integer spin particles, for example, an electron. (*Spin* is a quantum property of an elementary particle, its intrinsic angular momentum [1]. So, electrons, protons, and neutrinos have half-integer spin; the spin of photons and gravitons is an integer.).

The following properties are valid for the Pauli matrices.

(1) The Pauli matrices are Hermitian[1] and unitary (see Appendix B on page 103):

$$\forall k \in \{1, 2, 3\} \quad \sigma_k = \sigma_k^{\dagger} = \sigma_k^{-1}. \tag{2.2}$$

(2) $\forall k \in \{1, 2, 3\}$, the square of the Pauli matrix is equal to the identity matrix:

$$\sigma_i^2 = \begin{bmatrix} 1 & 0 \\ 0 & 1 \end{bmatrix}. \tag{2.3}$$

(3) $\forall i, j \in \{1, 2, 3\}$, the equalities

$$\sigma_i \sigma_j + \sigma_j \sigma_i = 2\delta_{ij} \begin{bmatrix} 1 & 0 \\ 0 & 1 \end{bmatrix} \tag{2.4}$$

are valid. Here, is used a notation for the Kronecker symbol (see definition on page 3).

Note that the value $\{A, B\} = AB + BA$ is usually called **anticommutator** of matrices A and B, and the value $[A, B] = AB - BA$ is called **commutator**. Next,

---

[1]Charles Hermite (1822–1901).

© The Author(s), under exclusive license to Springer Nature Switzerland AG 2021
S. Kurgalin and S. Borzunov, *Concise Guide to Quantum Computing*,
Texts in Computer Science, https://doi.org/10.1007/978-3-030-65052-0_2

for the identity matrix, we will apply the notation I, and for the zero one O. In particular, the formula (2.4) can be written as follows:

$$\{\sigma_i, \sigma_j\} = 2\delta_{ij}I. \tag{2.5}$$

The matrices A and B are called **commutative**, if $AB = BA$. Commutative matrices are always square and have the same order.

By definition, the condition $[A, B] = O$ is met for the commutative matrices [2]. It is clear that the Pauli matrices do not commutate with each other (see Exercise 1).

Example 2.1 Let us prove the **Jacobi**[2] **identity**

$$[[P, Q], R] + [[Q, R], P] + [[R, P], Q] \equiv O \tag{2.6}$$

that is valid for commutators of any matrices of size $n \times n$.
Proof.
We use the definition of the commutator $[P, Q] = PQ - QP$, then

$$[[P, Q], R] = [PQ - QP, R] = (PQ - QP)R - R(PQ - QP)$$
$$= PQR - QPR - RPQ + RQP. \tag{2.7}$$

Next, let us present in a similar manner the remaining summands in the sum:

$$[[Q, R], P] = QRP - RQP - PQR + PRQ, \tag{2.8}$$
$$[[R, P], Q] = RPQ - PRQ - QRP + QPR. \tag{2.9}$$

The sum of right-hand values (2.7), (2.8), and (2.9), as can be easily seen after the conversion of such summands, is zero. In this way, the Jacobi identity is proved. $\square$

Sometimes, in linear algebra and its applications, one has to use matrices split into rectangular parts or blocks [3,4]. Consider the rectangular matrix $A = (a_{ij})$, where $1 \leqslant i \leqslant m$, $1 \leqslant j \leqslant n$. Let $m = m_1 + m_2$ and $n = n_1 + n_2$.

Let us draw horizontal and vertical lines and split the matrix A into four rectangular blocks:

$$A = \left[\begin{array}{c|c} B_{11} & B_{12} \\ \hline B_{21} & B_{22} \end{array}\right] \begin{array}{l} \left.\vphantom{B_{11}}\right\} m_1 \\ \left.\vphantom{B_{21}}\right\} m_2 \end{array} \qquad \overbrace{n_1} \quad \overbrace{n_2} \tag{2.10}$$

---

[2] Carl Gustav Jacob Jacobi (1804–1851).

Thus, the matrix A is presented in the form of a **block matrix**, consisting of the blocks B_{11}, B_{12}, B_{21}, and B_{22} of size $m_1 \times n_1$, $m_1 \times n_2$, $m_2 \times n_1$, and $m_2 \times n_2$, respectively.

As an example of a block matrix setting, we provide the definition of the Dirac matrices. Four **Dirac matrices** α_1, α_2, α_3, and β are part of the equation named after him for a half-integer spin relativistic particle, and are expressed in terms of the Pauli matrices σ_k, $k = 1, 2, 3$, as follows [5]:

$$\alpha_k = \begin{bmatrix} O & \sigma_k \\ \sigma_k & O \end{bmatrix}, \quad \beta = \begin{bmatrix} I & O \\ O & -I \end{bmatrix}. \tag{2.11}$$

(*Relativistic particles* are the particles whose velocity is close to the velocity of light.)

Each of the Dirac matrices has a Hermitian property and a property of being unitary. Moreover, for all $l, m \in \{1, 2, 3\}$ the equalities

$$\alpha_l \alpha_m + \alpha_m \alpha_l = 2\delta_{lm} I, \tag{2.12}$$

$$\alpha_l \beta + \beta \alpha_l = O \tag{2.13}$$

are valid. Note that the size of matrices I and O in formulas (2.12) and (2.13) is equal to 4×4.

Using the concept of a block matrix, it is easy to write down the definition of the tensor product of square matrices $A \otimes B$, defined as $A = (a_{ij})$, $i, j = 1, 2, \ldots, n_A$, and $B = (b_{ij})$, $i, j = 1, 2, \ldots, n_B$:

$$A \otimes B - \begin{bmatrix} a_{11}B & a_{12}B & \ldots & a_{1n_A}B \\ u_{21}B & a_{22}B & \ldots & a_{2n_A}B \\ \vdots & \vdots & \ddots & \vdots \\ a_{n_A 1}B & a_{n_A 2}B & \ldots & a_{n_A n_A}B \end{bmatrix}, \tag{2.14}$$

where $a_{11}B$ is a block with size $n_B \times n_B$, consisting of elements of the form $a_{11}b_{ij}$, block $a_{12}B$ consists of elements of the form $a_{12}b_{ij}$, etc.

Note that for the tensor product operation, the sizes of matrices A and B do not have to be the same, i.e. $n_A \neq n_B$.

Example 2.2 Let us calculate tensor products $\sigma_1 \otimes \sigma_2$ and $\sigma_2 \otimes \sigma_1$.
Solution.
We use the definition of the tensor product:

$$\sigma_1 \otimes \sigma_2 = \begin{bmatrix} 0 & 1 \\ 1 & 0 \end{bmatrix} \otimes \begin{bmatrix} 0 & -i \\ i & 0 \end{bmatrix} = \begin{bmatrix} 0 \times \sigma_2 & 1 \times \sigma_2 \\ 1 \times \sigma_2 & 0 \times \sigma_2 \end{bmatrix} = \begin{bmatrix} 0 & 0 & 0 & -i \\ 0 & 0 & i & 0 \\ 0 & -i & 0 & 0 \\ i & 0 & 0 & 0 \end{bmatrix}. \tag{2.15}$$

In the same way, we obtain

$$\sigma_2 \otimes \sigma_1 = \begin{bmatrix} 0 & 0 & 0 & -i \\ 0 & 0 & -i & 0 \\ 0 & i & 0 & 0 \\ i & 0 & 0 & 0 \end{bmatrix}. \tag{2.16}$$

From the obtained results, it is easy to see the commutativity of the tensor product operation: $\sigma_1 \otimes \sigma_2 \neq \sigma_2 \otimes \sigma_1$. □

References

1. Landau LD, Lifshitz EM (1989) Quantum mechanics: non-relativistic theory. Course of theoretical physics, vol 3, 3rd edn. Pergamon Press, Oxford
2. Kurgalin S, Borzunov S (2020) Algebra and geometry with Python. Springer, Cham. xvi, 425 p
3. Gantmacher FR (2000) The theory of matrices, vol 1. American Mathematical Society, Providence, Rhode Island, x, 374 p
4. Magnus JR, Neudecker H (2019) Matrix differential calculus with applications in statistics and econometrics, 3rd edn. Wiley Series in Probability and Statistics. Wiley
5. Berestetskii VB, Lifshitz EM, Pitaevskii LP (1982) Quantum electrodynamics. Course of theoretical physics, vol 4, 2nd edn. Pergamon Press, Oxford

Basic Operations with Qubits

Consider basic operations with qubits.

A quantum gate effect on qubit $|\psi\rangle$ occurs by applying the **quantum-mechanical operator**, e.g. $U|\psi\rangle$ [1–3]. Operators can be represented as unitary matrices. In particular, the evolution of a single qubit is described by a unitary matrix of size 2×2.

The consistent application of a number of operators U_1, U_2, ..., U_n to one qubit is equivalent to the effect of some operator W in the form

$$W|\psi\rangle = U_n(U_{n-1}(\ldots(U_2(U_1|\psi\rangle))\ldots)) = (U_1 U_2 \ldots U_n)|\psi\rangle. \qquad (3.1)$$

Its matrix M_W is a product of matrices of U_i, $i = 1, 2, \ldots, n$, *in the reverse order* [4]:

$$M_W = M_{U_n} M_{U_{n-1}} \ldots M_{U_1}. \qquad (3.2)$$

Such an operator W is called a **product of operators** U_1, U_2, ..., U_n. Due to the noncommutativity of the matrix multiplication operation, the order in which quantum gates are applied is generally important.

Example 3.1 Let us show that the application of the operator σ_3 (see definition (2.1)) to a qubit in the state $|\psi\rangle = u|0\rangle + v|1\rangle$ brings it to the state $|\psi'\rangle = u|0\rangle - v|1\rangle$.

Proof.

Write a qubit $|\psi\rangle$ in matrix notation:

$$|\psi\rangle = \begin{bmatrix} u \\ v \end{bmatrix}. \qquad (3.3)$$

Let us define the action of σ_3 on this quantum state:

$$|\psi'\rangle = \sigma_3 \begin{bmatrix} u \\ v \end{bmatrix} = \begin{bmatrix} 1 & 0 \\ 0 & -1 \end{bmatrix} \begin{bmatrix} u \\ v \end{bmatrix} = \begin{bmatrix} u \\ -v \end{bmatrix} = u \begin{bmatrix} 1 \\ 0 \end{bmatrix} - v \begin{bmatrix} 0 \\ 1 \end{bmatrix} = u|0\rangle - v|1\rangle. \qquad (3.4)$$

$\square$

© The Author(s), under exclusive license to Springer Nature Switzerland AG 2021
S. Kurgalin and S. Borzunov, *Concise Guide to Quantum Computing*,
Texts in Computer Science, https://doi.org/10.1007/978-3-030-65052-0_3

The graphic representation of quantum operations in the form of circuits or diagrams (quantum circuits) is widely used.

A **quantum circuit**, or **network**, is an ordered sequence of elements and communication lines connecting them, i.e. wires. Usually, only **acyclic circuits** are considered in which data flow in one direction—from left to right, and the wires do not return bits to the previous position in the circuit. Input states are attributed to the wires coming from the left. In one time step, each wire may enter data in no more than one gate. The output states are read from the communication lines coming out of the circuit on the right.

As you can see, the circuit solves a problem of a fixed size. Unlike the quantum circuit, quantum algorithms are defined for input data of any size [5].

A quantum-mechanical operator U, which converts a one-qubit gate, is presented as follows:

$$|\psi_{\text{in}}\rangle \quad \boxed{U} \quad |\psi_{\text{out}}\rangle$$

The sequence of the quantum algorithm steps corresponds to the direction on the circuit from left to right.

Table 3.1 lists the frequently used gates that convert one qubit and the matrix notations of these gates.

Let us show the computation method of the quantum operation matrix based on its effect on basis vectors.

The **Hadamard**[1] **gate** converts the system state in accordance with the rule:

$$|0\rangle \rightarrow \frac{1}{\sqrt{2}}(|0\rangle + |1\rangle), \tag{3.5}$$

$$|1\rangle \rightarrow \frac{1}{\sqrt{2}}(|0\rangle - |1\rangle). \tag{3.6}$$

Consequently, the arbitrary state $|\psi\rangle$ will change in this case as follows:

$$|\psi\rangle = \begin{bmatrix} u \\ v \end{bmatrix} = u \begin{bmatrix} 1 \\ 0 \end{bmatrix} + v \begin{bmatrix} 0 \\ 1 \end{bmatrix}$$

$$\rightarrow u \frac{1}{\sqrt{2}}(|0\rangle + |1\rangle) + v \frac{1}{\sqrt{2}}(|0\rangle - |1\rangle) = \frac{1}{\sqrt{2}} \begin{bmatrix} 1 & 1 \\ 1 & -1 \end{bmatrix} \begin{bmatrix} u \\ v \end{bmatrix}. \tag{3.7}$$

Thus, the Hadamard gate, or the Hadamard element, corresponds to the matrix $\frac{1}{\sqrt{2}} \begin{bmatrix} 1 & 1 \\ 1 & -1 \end{bmatrix}$.

Of course, to execute complex algorithms, the qubits must interact with each other and exchange information. In this regard, logical operations involving two or more qubits are of special importance. In Table 3.2 are listed the most important gates that transform the state of two qubits.

Example 3.2 Let us define how the qubit $|\psi\rangle$ is converted under double application of the Hadamard gate:

---

[1]Jacques Salomon Hadamard (1865–1963).

Table 3.1 One qubit operations

Name	Notation	Matrix
Identity transformation	$-\boxed{I}-$	$I = \begin{bmatrix} 1 & 0 \\ 0 & 1 \end{bmatrix}$
Pauli Element X	$-\boxed{X}-$	$\sigma_1 = \begin{bmatrix} 0 & 1 \\ 1 & 0 \end{bmatrix}$
Pauli Element Y	$-\boxed{Y}-$	$\sigma_2 = \begin{bmatrix} 0 & -i \\ i & 0 \end{bmatrix}$
Pauli Element Z	$-\boxed{Z}-$	$\sigma_3 = \begin{bmatrix} 1 & 0 \\ 0 & -1 \end{bmatrix}$
Hadamard Element	$-\boxed{H}-$	$\frac{1}{\sqrt{2}} \begin{bmatrix} 1 & 1 \\ 1 & -1 \end{bmatrix}$
Phase Element	$-\boxed{S}-$	$\begin{bmatrix} 1 & 0 \\ 0 & i \end{bmatrix}$
Element $\pi/8$	$-\boxed{T}-$	$\begin{bmatrix} 1 & 0 \\ 0 & e^{i\pi/4} \end{bmatrix}$
Measurement	$-\boxed{\angle}$	Projection on the state of computational basis

Table 3.2 Two qubits operations

Name	Notation	Matrix
Exchange		$\begin{bmatrix} 1 & 0 & 0 & 0 \\ 0 & 0 & 1 & 0 \\ 0 & 1 & 0 & 0 \\ 0 & 0 & 0 & 1 \end{bmatrix}$
Controlled NOT		$\begin{bmatrix} 1 & 0 & 0 & 0 \\ 0 & 1 & 0 & 0 \\ 0 & 0 & 0 & 1 \\ 0 & 0 & 1 & 0 \end{bmatrix}$
Controlled Phase Element		$\begin{bmatrix} 1 & 0 & 0 & 0 \\ 0 & 1 & 0 & 0 \\ 0 & 0 & 1 & 0 \\ 0 & 0 & 0 & i \end{bmatrix}$

$$|\psi\rangle \quad \boxed{H} \qquad \boxed{H} \qquad |\psi'\rangle$$

Solution.

As shown above, in the matrix representation, the Hadamard element is described by the matrix

$$M_H = \frac{1}{\sqrt{2}} \begin{bmatrix} 1 & 1 \\ 1 & -1 \end{bmatrix}. \tag{3.8}$$

Let us compute a matrix corresponding to the double application of the Hadamard element as a matrix product (see formula (3.1)):

$$M_H M_H = \frac{1}{\sqrt{2}} \begin{bmatrix} 1 & 1 \\ 1 & -1 \end{bmatrix} \times \frac{1}{\sqrt{2}} \begin{bmatrix} 1 & 1 \\ 1 & -1 \end{bmatrix} = \frac{1}{2} \begin{bmatrix} 2 & 0 \\ 0 & 2 \end{bmatrix} = \begin{bmatrix} 1 & 0 \\ 0 & 1 \end{bmatrix}. \tag{3.9}$$

The identity matrix is obtained, therefore a double application of the Hadamard gate returns the qubit to its initial state. $\qquad\square$

As already noted on page 4, the qubit **measurement** operation is fundamental. This operation is performed with a classic (not a quantum) instrument, and it converts the state of one qubit $|\psi\rangle = u\,|0\rangle + v\,|1\rangle$ into a probabilistic classic bit M. According to quantum theory principles, M equals zero with a probability of $|u|^2$ and one with a probability of $|v|^2$. Specifically, the results of qubit measurements allow the use of quantum-mechanical circuits to obtain information in the macro world.

Symbolically, the measurement result is represented in the circuit by a double line:

$$|\psi\rangle \quad \boxed{\measuredangle}$$

Example 3.3 The following quantum state is set:

$$|\psi\rangle = \frac{1}{\sqrt{2}}\,|0\rangle + \frac{1}{\sqrt{2}}e^{i\varphi}\,|1\rangle, \tag{3.10}$$

where φ is a real number. Let us define the result of measuring this state in bases $B_1 = \{|0\rangle, |1\rangle\}$ and $B_2 = \{|+\rangle, |-\rangle\}$, where $|+\rangle = \frac{1}{2}(|0\rangle + |1\rangle)$ and $|-\rangle = \frac{1}{2}(|0\rangle - |1\rangle)$.

Solution.

First of all, we consider the basis B_1. According to the measurement postulate (see Appendix A on page 105), the probabilities of getting 0 or 1 as a result of the measurement of the state $u\,|0\rangle + v\,|1\rangle$ is equal to $|u|^2$ and $|v|^2$, respectively. In our example, $u = \frac{1}{\sqrt{2}}$ and $v = \frac{1}{\sqrt{2}}e^{i\varphi}$, therefore the result of the measurement of M is

$$M = \begin{cases} 0 \text{ with a probability of } p_0 = \dfrac{1}{2}, \\ 1 \text{ with a probability of } p_1 = \dfrac{1}{2}. \end{cases} \tag{3.11}$$

Note that information about the value of φ using a measurement in the basis B_1 cannot be obtained.

Let us turn to measurement using the second basis $B_2 = \{|+\rangle, |-\rangle\}$. We express the vectors of the computational basis through $|+\rangle$ and $|-\rangle\}$:

$$|0\rangle = \frac{1}{\sqrt{2}}(|+\rangle + |-\rangle),$$

$$|1\rangle = \frac{1}{\sqrt{2}}(|+\rangle + |-\rangle). \tag{3.12}$$

In this regard, we obtain

$$|\psi\rangle = \frac{1}{\sqrt{2}}|0\rangle + \frac{1}{\sqrt{2}}e^{i\varphi}|1\rangle$$

$$= \frac{1}{2}(|+\rangle + |-\rangle) + \frac{1}{2}e^{i\varphi}(|+\rangle + |-\rangle)$$

$$= \frac{1}{2}(1 + e^{i\varphi})|+\rangle + \frac{1}{2}(1 - e^{i\varphi})|-\rangle. \tag{3.13}$$

The probability of finding the system in the state $|+\rangle$ is

$$p_+ = \frac{1}{2}\left(1 + e^{i\varphi}\right) \times \left(\frac{1}{2}(1 + e^{i\varphi})\right)^*$$

$$= \frac{1}{4}\left(1 + e^{i\varphi}\right)\left(1 + e^{-i\varphi}\right) = \frac{1}{4}\left(2 + e^{i\varphi} + e^{-i\varphi}\right). \tag{3.14}$$

Using the **Euler**[2] **formula** $e^{i\varphi} = \cos\varphi + i\sin\varphi$, that is known from the course of mathematical analysis, the expression for p_+ can be simplified as:

$$p_+ = \frac{1}{2}(1 + \cos\varphi) = \cos^2(\varphi/2). \tag{3.15}$$

In the same way, we get the probability of finding the system in the state $|-\rangle$:

$$p_- = \frac{1}{2}(1 - \cos\varphi) = \sin^2(\varphi/2). \tag{3.16}$$

In accordance with the law of conservation of total probability, the equality $p_+ + p_- = 1$ is fulfilled.

It should be noted that measurement in the basis B_2 allowed to save information about the value φ. □

Let us consider more complex, two-qubit gates. One of the most important of such elements is the element Controlled NOT (CNOT). It has two input qubits which are called the **control** and the **controlled** one. The controlled qubit is also known as the **target** qubit.

The circuit representation for the CNOT gate is shown below.

The upper communication line represents the control qubit, and the lower line represents the target qubit. CNOT functions according to the following rule:

---

[2]Leonhard Euler (1707–1783).

- if the control qubit is $|0\rangle$, the target qubit stays unchanged;
- if the control qubit is set to $|1\rangle$, the value of the target qubit is inverted.

As a result, a NOT operation is performed with the target qubit when the first—control—qubit is in the state $|1\rangle$.

The specified rule of state conversion corresponds to the matrix of the CNOT operator

$$U_{\text{CNOT}} = \begin{bmatrix} 1 & 0 & 0 & 0 \\ 0 & 1 & 0 & 0 \\ 0 & 0 & 0 & 1 \\ 0 & 0 & 1 & 0 \end{bmatrix}. \tag{3.17}$$

There is another way to present the effect of a CNOT gate. Let us look at the state of two qubits $|\psi_1\psi_2\rangle$, applied to the input of this element. According to the transformation (3.17), the CNOT gate performs the transformation of $|\psi_1, \psi_2\rangle \rightarrow |\psi_1, \psi_1 \oplus \psi_2\rangle$. In other words, the control and controlled qubits are added modulo two and the result is saved in the controlled qubit.

Following from (3.17), reapplying the CNOT gate returns the two-qubit system to its initial state: the matrix product $U_{\text{CNOT}} \times U_{\text{CNOT}}$ equals the identity matrix of size 4×4.

Example 3.4 Let us find a matrix representation of the following quantum circuit:

Solution. The quantum circuit consists of two elements "Controlled NOT", or as it is also called, "CNOT".

The matrix of the CNOT element has the form (see Table 3.2):

$$\begin{bmatrix} 1 & 0 & 0 & 0 \\ 0 & 1 & 0 & 0 \\ 0 & 0 & 0 & 1 \\ 0 & 0 & 1 & 0 \end{bmatrix}. \tag{3.18}$$

Taking into account the CNOT matrix representation, let us calculate how the arbitrary state $|\psi_1\psi_2\rangle = (u_1 |0\rangle + v_1 |1\rangle)(u_2 |0\rangle + v_2 |1\rangle) = a_0 |00\rangle + a_1 |01\rangle + a_2 |10\rangle + a_3 |11\rangle$ will change after the action of the first CNOT:

$$|\psi_1\psi_2\rangle = \begin{bmatrix} a_0 \\ a_1 \\ a_2 \\ a_3 \end{bmatrix} \rightarrow \begin{bmatrix} 1 & 0 & 0 & 0 \\ 0 & 1 & 0 & 0 \\ 0 & 0 & 0 & 1 \\ 0 & 0 & 1 & 0 \end{bmatrix} \begin{bmatrix} a_0 \\ a_1 \\ a_2 \\ a_3 \end{bmatrix} = \begin{bmatrix} a_0 \\ a_1 \\ a_3 \\ a_2 \end{bmatrix}. \tag{3.19}$$

This is equivalent to the fact that the states of computational basis (1.5) are converted according to the rule:

$$
\begin{cases}
|00\rangle \rightarrow |00\rangle, \\
|01\rangle \rightarrow |01\rangle, \\
|10\rangle \rightarrow |11\rangle, \\
|11\rangle \rightarrow |10\rangle.
\end{cases}
\tag{3.20}
$$

Note that the next CNOT gate takes the input states in the order opposite to the first element. The rule of basis state conversion in this case is as follows:

$$
\begin{cases}
|00\rangle \rightarrow |00\rangle, \\
|01\rangle \rightarrow |11\rangle, \\
|10\rangle \rightarrow |10\rangle, \\
|11\rangle \rightarrow |01\rangle.
\end{cases}
\tag{3.21}
$$

Therefore, the next stage of the quantum system evolution is described by the matrix

$$
\begin{bmatrix}
1 & 0 & 0 & 0 \\
0 & 0 & 0 & 1 \\
0 & 0 & 1 & 0 \\
0 & 1 & 0 & 0
\end{bmatrix}.
\tag{3.22}
$$

Perform matrix computations:

$$
\begin{bmatrix} a_0 \\ a_1 \\ a_3 \\ a_2 \end{bmatrix} \rightarrow
\begin{bmatrix}
1 & 0 & 0 & 0 \\
0 & 0 & 0 & 1 \\
0 & 0 & 1 & 0 \\
0 & 1 & 0 & 0
\end{bmatrix}
\begin{bmatrix} a_0 \\ a_1 \\ a_3 \\ a_2 \end{bmatrix} =
\begin{bmatrix} a_0 \\ a_2 \\ a_3 \\ a_1 \end{bmatrix}.
\tag{3.23}
$$

As a result, the initial state $|\psi_1\psi_2\rangle = [a_0, a_1, a_2, a_3]^T$ will transfer to $[a_0, a_2, a_3, a_1]^T$. The matrix representation of the analyzed circuit can be written in the following form:

$$
M_W =
\begin{bmatrix}
1 & 0 & 0 & 0 \\
0 & 0 & 0 & 1 \\
0 & 0 & 1 & 0 \\
0 & 1 & 0 & 0
\end{bmatrix}
\begin{bmatrix}
1 & 0 & 0 & 0 \\
0 & 1 & 0 & 0 \\
0 & 0 & 0 & 1 \\
0 & 0 & 1 & 0
\end{bmatrix} =
\begin{bmatrix}
1 & 0 & 0 & 0 \\
0 & 0 & 1 & 0 \\
0 & 0 & 0 & 1 \\
0 & 1 & 0 & 0
\end{bmatrix}.
\tag{3.24}
$$

$\square$

Example 3.5 Consider the transformation to which corresponds the following quantum circuit:

Input state is equal to $|11\rangle$. Determine the state at the output of this circuit.

Solution.
After applying the first Hadamard transformation to the upper qubit we obtain

$$|11\rangle \to 1/\sqrt{2}(|01\rangle - |11\rangle). \tag{3.25}$$

Right after the CNOT action, we obtain

$$1/\sqrt{2}(|01\rangle - |11\rangle) \to 1/\sqrt{2}(|01\rangle - |10\rangle). \tag{3.26}$$

Then there is performed the repeated Hadamard transformation to the upper qubit:

$$1/\sqrt{2}(|01\rangle - |10\rangle) \to 1/\sqrt{2}((|01\rangle + |11\rangle)/\sqrt{2} - (|00\rangle - |10\rangle)/\sqrt{2})$$
$$= (-|00\rangle + |01\rangle + |10\rangle + |11\rangle)/2. \tag{3.27}$$

So we get the final result: the state of the circuit output is equal to $(-|00\rangle + |01\rangle + |10\rangle + |11\rangle)/2$.

Theorem of universal set of elements. *Over a system consisting of n qubits, using only one-qubit elements and CNOT elements it is possible to implement an arbitrary unitary operation.*

The proof of this important result is presented in works [6–8].

References

1. Steane A (1988) Quantum computing. Rep Prog Phys 61(2):117–173
2. Williams CP (2011) Explorations in quantum computing, 2nd edn. (Series: Texts in computer science). Springer, xxii, 717 p
3. Zapryagaev SA (2015) Introduction to quantum information systems [in Russian] (Series: Voronezh state university textbook). VSU Publishing House, Voronezh, 219 p
4. Nielsen MA, Chuang IL (2010) Quantum computation and quantum information. 10th anniversary edition. Cambridge University Press, Cambridge, xxxi, 676 p
5. Kitaev AYu, Shen AH, Vyalyi MN (2002) Classical and quantum computation (Series: Graduate studies in mathematics, vol 47). American Mathematical Society, 257 p
6. Deutsch D (1995) Universality in quantum computation. Proc R Soc Lond Ser A 449:669–677
7. DiVincenzo DP (1995) Two-bit gates are universal for quantum computation. Phys Rev A 50(2):1015
8. Lloyd S (1995) Almost any quantum logic gate is universal. Phys Rev Lett 75(2):346–349

Entangled States

4

Let us consider an important case of a two-qubit system state, whose wave function is equal to

$$|\psi\rangle = \frac{1}{\sqrt{2}}(|00\rangle + |11\rangle). \tag{4.1}$$

This state is called the **Bell**[1] **state**. In this state, when measuring the first qubit, two results are possible:

- $M = 0$ with the probability $p = 1/2$, and final state $|\psi'\rangle = |00\rangle$;
- $M = 1$ with the same probability $p = 1/2$, and final state $|\psi'\rangle = |11\rangle$.

Measuring the second qubit always gives the same result as measuring the first qubit.

To prepare the state (4.1), one can use the following quantum circuit:

At the first stage of this circuit, the Hadamard transformation converts the upper qubit into a superposition state $(|0\rangle + |1\rangle)/\sqrt{2}$. At the next stage, this superposition moves to the control input of the CNOT gate, and the target qubit is inverted if and only if the control qubit value is equal to one.

Using the matrix approach, we determine the state at the output of the circuit. At the input, we have

$$|\psi\rangle = |0\rangle\,|0\rangle = \begin{bmatrix} 1 \\ 0 \\ 0 \\ 0 \end{bmatrix}. \tag{4.2}$$

---

[1] John Stewart Bell (1928–1990).

© The Author(s), under exclusive license to Springer Nature Switzerland AG 2021
S. Kurgalin and S. Borzunov, *Concise Guide to Quantum Computing*,
Texts in Computer Science, https://doi.org/10.1007/978-3-030-65052-0_4

Let us apply the Hadamard transformation to the first qubit:

$$|\psi\rangle \to U_H \times \begin{bmatrix} 1 \\ 0 \end{bmatrix} = \frac{1}{\sqrt{2}} \begin{bmatrix} 1 & 1 \\ 1 & -1 \end{bmatrix} \times \begin{bmatrix} 1 \\ 0 \end{bmatrix} = \frac{1}{\sqrt{2}} \begin{bmatrix} 1 \\ 1 \end{bmatrix}. \tag{4.3}$$

The state of the second qubit has not yet changed and is described by the vector $\begin{bmatrix} 1 \\ 0 \end{bmatrix}$.
The vector of the whole system is

$$|\psi_1\rangle = \begin{bmatrix} \frac{1}{\sqrt{2}} \\ 0 \\ \frac{1}{\sqrt{2}} \\ 0 \end{bmatrix}. \tag{4.4}$$

Then we calculate the result of the controlled NOT:

$$|\psi_1\rangle \to |\psi'\rangle = U_{\text{CNOT}} \begin{bmatrix} 1/\sqrt{2} \\ 0 \\ 1/\sqrt{2} \\ 0 \end{bmatrix} = \begin{bmatrix} 1 & 0 & 0 & 0 \\ 0 & 1 & 0 & 0 \\ 0 & 0 & 0 & 1 \\ 0 & 0 & 1 & 0 \end{bmatrix} \times \begin{bmatrix} 1/\sqrt{2} \\ 0 \\ 1/\sqrt{2} \\ 0 \end{bmatrix} = \begin{bmatrix} 1/\sqrt{2} \\ 0 \\ 0 \\ 1/\sqrt{2} \end{bmatrix}. \tag{4.5}$$

We obtain the wave function $|\psi'\rangle = (|00\rangle + |11\rangle)/\sqrt{2}$.

Note. The set consisting of four wave functions

$$|\beta_1\rangle = \frac{1}{\sqrt{2}}(|00\rangle + |11\rangle),$$

$$|\beta_2\rangle = \frac{1}{\sqrt{2}}(|01\rangle + |10\rangle),$$

$$|\beta_3\rangle = \frac{1}{\sqrt{2}}(|00\rangle - |11\rangle),$$

$$|\beta_4\rangle = \frac{1}{\sqrt{2}}(|01\rangle - |10\rangle) \tag{4.6}$$

is called the **Bell basis**. Elements $|\beta_1\rangle$–$|\beta_4\rangle$ of this set are hence called **Bell states**. By direct checking (see Exercise 56), it is easy to make sure that they form an orthonormalized basis.

The Bell state $|\beta_1\rangle = \frac{1}{\sqrt{2}}(|00\rangle + |11\rangle)$ is a simple example of entangled states.

Entangled states are those states of the quantum system in which qubits interact with each other, and a description of its state with the product of wave functions of independent qubits is impossible.

We prove the entanglement of the state $\psi = (|00\rangle + |11\rangle)/\sqrt{2}$ as follows. Suppose that there is a possibility to represent ψ as a product of wave functions of non-interacting qubits:

$$\exists u_1, v_1, u_2, v_2 \in \mathbb{C}: \quad |\psi\rangle = (u_1 |0\rangle + v_1 |1\rangle)(u_2 |0\rangle + v_2 |1\rangle). \tag{4.7}$$

The complex parameters u_1, v_1, u_2, and v_2 have the meaning of amplitudes in the superposition of basis states. Let us set the constraints to which these parameters must satisfy. To do this, remove the parentheses in (4.7):

$$|\psi\rangle = u_1 u_2 |0\rangle |0\rangle + u_1 v_2 |0\rangle |1\rangle + u_2 v_1 |1\rangle |0\rangle + v_1 v_2 |1\rangle |1\rangle . \qquad (4.8)$$

We obtain a system of equations regarding complex values u_1, v_1, u_2, and v_2:

$$\begin{cases} u_1 u_2 = \dfrac{1}{\sqrt{2}}, \\ u_1 v_2 = \quad 0, \\ u_2 v_1 = \quad 0, \\ u_2 v_2 = \dfrac{1}{\sqrt{2}}. \end{cases} \qquad (4.9)$$

Since the second equation of the system $u_1 v_2 = 0$ causes $u_1 u_2 = 0$ or $v_1 v_2 = 0$, the system is incompatible and the set of its solutions is empty. Consequently, there are no such wave functions whose product corresponds to the state $|\psi\rangle$. This proves the entanglement of the Bell state.

In entangled states, there are manifested correlations that do not have classical analogues.

The introduced Bell state above is sometimes called the **EPR pair** after Einstein,[2] Podolsky[3], and Rosen,[4] who studied similar states at the stage of the formation of quantum theory [1–4].

In a mental experiment, considered in [1], there was analyzed the decomposition of some particle c into two identical particles a and b. According to the law of momentum conservation of an isolated quantum-mechanical system, the impulses of the formed particles p_a and p_b are correlated with the impulse of the initial p_c by the following relation:

$$p_a + p_b = p_c. \qquad (4.10)$$

Note that the states $|a\rangle$ and $|b\rangle$ are entangled and form an EPR pair.

Let us measure the impulse of the particle a. Thus, it is possible to calculate the impulse of another particle b without performing the procedure of its interaction with the classical instrument: $p_b = p_c - p_a$. Having further measured the coordinate of the particle b, we obtain for this particle certain values of both impulses and coordinates, which contradicts the bases of quantum mechanics, in particular, the Heisenberg[5] uncertainty principle. Moreover, in this mental experiment, measuring the impulse of one of the particles is equivalent to measuring the impulse of the other. In return, from the point of view of the authors of the paper [1], it contradicts the general physical principle of causality.

In the described EPR experiment, after measuring the impulse p_a, the second particle goes to the state with a certain impulse. Measuring the coordinate of a particle

---

[2]Albert Einstein (1879–1955).
[3]Boris Yakovlevich Podolsky (1896–1966).
[4]Nathan Rosen (1909–1995).
[5]Werner Karl Heisenberg (1901–1976).

b will lead to the fact that as a result of the interaction with the classical instrument, its impulse, in general, will change. Consequently, the simultaneous measurement of the coordinate and the impulse in this experiment is not performed.

The second particle b gets a certain impulse immediately after the measurement made over the particle a, even if the particles are spatially separated by a large distance. It follows that quantum theory has the property of **nonlocality** [5,6].

No-cloning theorem. *It is impossible to create an exact copy of an arbitrary unknown quantum state.*

The reasoning will follow the paper [7] (see also [8]). We use the method of proving "by contradiction".

Suppose we need to create an exact copy of the system S that is in the state $|\psi\rangle$. We denote the backup system that will be used to record and store the state $|\psi\rangle$ by T; let it initially be in state $|\Lambda\rangle$.

The system is in general, as easy to see, in a state with the wave function

$$|\psi\rangle\,|\Lambda\rangle \tag{4.11}$$

Suppose there is a unitary transformation U_X that allows us to write in $|\Lambda\rangle$ an exact copy of the initial state $|\psi\rangle$:

$$|\psi\rangle\,|\Lambda\rangle \to U_X\,|\psi\rangle\,|\Lambda\rangle = |\psi\rangle\,|\psi\rangle. \tag{4.12}$$

Let us choose two arbitrary states of the system S, for example, $|\psi'\rangle$ and $|\psi''\rangle$. The action of a unitary transformation U_X in these cases can be represented as follows:

$$\begin{aligned}
|\psi'\rangle\,|\Lambda\rangle &\to |\psi'\rangle\,|\psi'\rangle, \\
|\psi''\rangle\,|\Lambda\rangle &\to |\psi''\rangle\,|\psi''\rangle.
\end{aligned} \tag{4.13}$$

The main idea to prove the no-cloning theorem is to apply U_X to the superposition $u\,|\psi'\rangle + v\,|\psi''\rangle$. Due to the linearity of quantum-mechanical operations (see Appendix A on page 105),

$$|\psi\rangle\,|\Lambda\rangle \to u\,|\psi'\rangle\,|\psi'\rangle + v\,|\psi''\rangle\,|\psi''\rangle. \tag{4.14}$$

But the result required is different from (4.14):

$$\begin{aligned}
|\psi\rangle\,|\Lambda\rangle &\to |\psi\rangle\,|\psi\rangle \\
&= (u\,|\psi'\rangle + v\,|\psi''\rangle)(u\,|\psi'\rangle + v\,|\psi''\rangle) \\
&= u^2\,|\psi'\rangle\,|\psi'\rangle + uv(|\psi'\rangle\,|\psi''\rangle + |\psi'\rangle\,|\psi''\rangle) + v^2\,|\psi''\rangle\,|\psi''\rangle.
\end{aligned} \tag{4.15}$$

Thus, the assumption that the operation U_X can clone an arbitrary quantum state is not correct. The no-cloning theorem is proved.

Note that although it is impossible to create copies of unknown arbitrary quantum states, a copy of known states can often be obtained quite easily. For example, knowing that a qubit is converted to the state $|0\rangle$ or $|1\rangle$, measuring this state will give one classic bit of information $M = 0$ or $M = 1$. Using M, it is possible to form copies of the state $|M\rangle$ based on ancillary qubits.

Example 4.1 The set of qubits is brought to the state

$$|\psi\rangle = \frac{1}{3}(|001\rangle + |010\rangle + \sqrt{7}\,|100\rangle).\tag{4.16}$$

With what probability will the result of the measurement of the first qubit be $|1\rangle$?
Solution.
Let us write the decomposition $|\psi\rangle$ on a computational basis:

$$\begin{aligned}|\psi\rangle = {}& a_0\,|000\rangle + a_1\,|001\rangle + a_2\,|010\rangle + a_3\,|011\rangle \\ &+ a_4\,|100\rangle + a_5\,|101\rangle + a_6\,|110\rangle + a_7\,|111\rangle.\end{aligned}\tag{4.17}$$

The result of the measurement of the first qubit, equal to $|1\rangle$, will be obtained with
a probability $p = |a_4|^2 + |a_5|^2 + |a_6|^2 + |a_7|^2$. Taking into account the equalities
$a_4 = \sqrt{7}/3$ and $a_5 = a_6 = a_7 = 0$, we obtain $p = 7/9$. □

It is proved that the entanglement of a quantum state is not a necessary condition
for the possibility of conducting quantum calculations with its help. Using unentan-
gled systems for calculations, many quantum algorithms, however, demonstrate less
asymptotic complexity in comparison with classical methods of solution [9,10].

References

1. Einstein A, Podolsky B, Rosen N (1935) Can quantum mechanical description of physical
   reality be considered complete? Phys Rev 47:777–780
2. Bohr N (1935) Can quantum-mechanical description of physical reality be considered complete? Phys Rev 48:696–702
3. Furry WH (1936) Note on the quantum-mechanical theory of measurement. Phys Rev 49:393–
   399
4. Furry WH (1936) Remark on measurements in quantum theory. Phys Rev 49:476
5. Uola R, Costa ACS, Nguyen HC, Gühne O (2020) Quantum steering. Rev Modern Phys
   92(1):015001-1–015001-40
6. Wharton KB, Argaman N (2020) Colloquium: Bell's theorem and locally mediated reformu-
   lations of quantum mechanics. Rev Modern Phys 92(2):021002-1–021002-23
7. Wootters WK, Zurek WH (1982) A single quantum cannot be cloned. Nature 299:802–803
8. Dieks D (1982) Communication by EPR devices. Phys Lett A 92(6):271–272
9. Biham E, Brassard G, Kenigsberg D, Mor T (2004) Quantum computing without entanglement.
   Theor Comput Sci 320:15–33
10. Horodecki R, Horodecki P, Horodecki M, Horodecki K (2009) Quantum entanglement. Rev
   Modern Phys 81(2):865–942

Quantum Teleportation

5

Suppose that two physical laboratories, denoted by A and B, are conducting research in the field of quantum computing. There is an important result that shows the following: it is possible to transfer from lab A to lab B an arbitrary qubit $|\psi\rangle = u\,|0\rangle + v\,|1\rangle$, where $u, v \in \mathbb{C}$, using only a *classical* communication channel and a preset shared EPR pair.

Now we show how to realize it. The initial state of the quantum system is as follows:

$$(u\,|0\rangle + v\,|1\rangle)(|00\rangle + |11\rangle)/\sqrt{2}. \tag{5.1}$$

To be specific, we will assume that the first qubit in the EPR pair is in lab A, the second one is in lab B. Note that the coefficients u and v are unknown to every one of the researchers.

To transfer information, the CNOT operation over the two qubits in A, which are combined using a curly bracket in the scheme below, is performed. Then, the Hadamard transformation on the first qubit is performed.

At the next stage, two qubits in A are measured in a computational basis. The measurement results form two classic bits, which are sent to B.

In lab B, an unknown state $|\psi\rangle$ is restored based on information in the form of classic bits using the following scheme [1,2]:

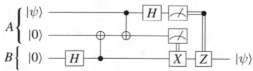

Thus, the quantum state has been transferred from one laboratory to another in the absence of a quantum channel between the sender and the recipient, i.e. **quantum teleportation** is performed.

Note that quantum teleportation does not contradict the theorem about the impossibility of copying an arbitrary state, because in laboratory A the state $|\psi\rangle$ will be lost and the corresponding qubit will be converted to the state $|0\rangle$ or $|1\rangle$.

S. Kurgalin and S. Borzunov, *Concise Guide to Quantum Computing*,
Texts in Computer Science, https://doi.org/10.1007/978-3-030-65052-0_5

Example 5.1 The physical laboratory A has prepared a two-qubit state

$$|\psi\rangle = u\,|0\rangle + v\,|1\rangle \tag{5.2}$$

with some complex coefficients u and v that meet the normalization condition $|u|^2 + |v|^2 = 1$. The Hadamard operator was applied to the first qubit and then it was measured. Let us define the state of the second qubit after the mentioned actions.

Solution.

The action of the Hadamard operator will put the system in the state

$$|\psi'\rangle = \frac{1}{\sqrt{2}}(u\,|00\rangle + v\,|01\rangle + v\,|10\rangle - v\,|11\rangle). \tag{5.3}$$

Next, consider two cases.

1. The result of measuring the first qubit $M = 0$.

Then the system state after measurement is described by the wave function $|0\rangle\,(u\,|0\rangle + v\,|1\rangle)$. Therefore, the second qubit is in the state $|0\rangle\,(u\,|0\rangle + v\,|1\rangle)$.

2. The result of measuring the first qubit $M = 1$.

In this case, the second qubit is in the state $(u\,|0\rangle - v\,|1\rangle)$.

As a result, let us write the final answer: the state of the second qubit after measuring the first one is $(u\,|0\rangle + (-1)^M v\,|1\rangle)$.   □

References

1. Bennett CH, Brassard G, Crépeau C et al (1993) Teleporting an unknown quantum state via dual classical and Einstein–Podolsky–Rosen channels. Phys Rev Lett 70(13):1895–1899
2. Nielsen MA, Chuang IL (2010) Quantum computation and quantum information, 10th anniversary edition. Cambridge University Press, Cambridge, xxxi, 676 p

Universal Elements of Quantum Circuits

6

We will need some definitions from Boolean algebra for further description. Recall the main definitions [1,2].

A **Boolean**[1] **variable** p can take the values 0 or 1 (and only them), $p \in \mathbb{B} \equiv \{0, 1\}$.

Boolean algebra is a set $\mathbb{B}$ with operations of disjunction ($\vee$), conjunction ($\wedge$), and negation ($\overline{}$) defined on it.

The effect of the main operations on the Boolean variables is presented in Table 6.1.

A **Boolean expression** is constructed from Boolean variables using the operations $\vee$, $\wedge$, $\overline{}$, and brackets. The expression of the form $x_1 \wedge x_2$ is sometimes written as $x_1 \,\&\, x_2$ or simply $x_1 x_2$.

The collection $\mathbf{x} = (x_1, x_2, \ldots, x_n)$, where $x_i \in \{0, 1\}$, $1 \leqslant i \leqslant n$, is called the **Boolean tuple (vector)**. The elements of the tuple are **components**, or **coordinates**. The function $f \colon \mathbb{B}^n \to \mathbb{B}$, where $f(x_1, \ldots, x_n)$ is a Boolean expression, is called the **Boolean function**.

One of the ways to define a Boolean function $f(\mathbf{x})$ is by the table of values (Table 6.2). This table is also called **truth table**. As is easy to see, the number of rows in the table is 2^n, where n is the number of components of the vector $\mathbf{x}$.

There is also used a method of defining a function with a vector of values $\boldsymbol{\alpha} = (\alpha_0, \alpha_1, \ldots, \alpha_{2^n-1})$, where the coordinate α_i is equal to the value of the function on the tuple of variables located in the ith row, $0 \leqslant i \leqslant 2^n - 1$.

Table 6.3 lists all functions from $P_2(2)$. The symbols "$\downarrow$", "$|$", "$\oplus$", "$\leftrightarrow$", and "$\to$" are notations for combinations of disjunction, conjunction, and negation.

The functions listed in Table 6.3 are widespread and many of them have special names, for example: Peirce's arrow ($\downarrow$), Sheffer stroke ($|$), addition modulo two ($\oplus$), logical equivalence ($\leftrightarrow$), and implication ($\to$).

---

[1] George Boole (1815–1864).

© The Author(s), under exclusive license to Springer Nature Switzerland AG 2021
S. Kurgalin and S. Borzunov, *Concise Guide to Quantum Computing*,
Texts in Computer Science, https://doi.org/10.1007/978-3-030-65052-0_6

Table 6.1 Negation, disjunction, and conjunction

		p	q	$p \vee q$	$p \wedge q$
p	$\overline{p}$	0	0	0	0
0	1	0	1	1	0
1	0	1	0	1	0
		1.	1	1	1

Table 6.2 Table of values of function $f(\mathbf{x})$

x_1	x_2	$\ldots$	x_{n-1}	x_n	$f(x_1, x_2, \ldots, x_{n-1}, x_n)$
0	0	$\ldots$	0	0	$f(0, 0, \ldots, 0, 0)$
0	0	$\ldots$	0	1	$f(0, 0, \ldots, 0, 1)$
0	0	$\ldots$	1	0	$f(0, 0, \ldots, 1, 0)$
1	1	$\ldots$	1	1	$f(1, 1, \ldots, 1, 1)$

Elementary conjunction, or **minterm**, is a conjunction of n various Boolean variables or their negations $x_1^{\sigma_1} \wedge x_2^{\sigma_2} \wedge \ldots \wedge x_n^{\sigma_n}$, where $n \geqslant 1$ and

$$x_i^{\sigma_i} = \begin{cases} \overline{x}_i, & \sigma_i = 0, \\ x_i, & \sigma_i = 1. \end{cases} \tag{6.1}$$

The number of variables is called the **rank** of an elementary conjunction. Elementary conjunction is called **monotone**, if it contains no negations of variables. Elementary conjunction takes a value equal to 1 on only one set of arguments' values. Therefore, an arbitrary function that is not identically equal to zero can be represented as a disjunction of elementary conjunctions. An expression of the form

$$\mathcal{D} = K_1 \vee K_2 \vee \ldots \vee K_l, \tag{6.2}$$

where K_i, $i = 1, 2, \ldots, l$ are pairwise different elementary conjunctions, is called **disjunctive normal form** (DNF); here l is the length of DNF.

Example 6.1 Let us build the perfect disjunctive normal form of the function defined by the vector of values $\boldsymbol{\alpha}_f = (1011\ 0011)$.

Solution.

The function $f(x_1, x_2, x_3)$ takes value equal to 1 on the following tuples of arguments (x_1, x_2, x_3):

$$(0, 0, 0), \ (0, 1, 0), \ (0, 1, 1), \ (1, 1, 0), \ (1, 1, 1).$$

We combine the corresponding elementary conjunctions and obtain the perfect disjunctive normal form:

$$\mathcal{D}_f = \overline{x}_1 \overline{x}_2 \overline{x}_3 \vee \overline{x}_1 x_2 \overline{x}_3 \vee \overline{x}_1 x_2 x_3 \vee x_1 x_2 \overline{x}_3 \vee x_1 x_2 x_3. \tag{6.3}$$

$\square$

Table 6.3 Functions of two arguments

Notation		0	$x_1 \wedge x_2$	$\overline{(x_1 \rightarrow x_2)}$	x_1
x_1	x_2	f_0	f_1	f_2	f_3
0	0	0	0	0	0
0	1	0	0	0	0
1	0	0	0	1	1
1	1	0	1	0	1

Notation		$\overline{(x_2 \rightarrow x_1)}$	x_2	$x_1 \oplus x_2$	$x_1 \vee x_2$
x_1	x_2	f_4	f_5	f_6	f_7
0	0	0	0	0	0
0	1	1	1	1	1
1	0	0	0	1	1
1	1	0	1	0	1

Notation		$x_1 \downarrow x_2$	$x_1 \leftrightarrow x_2$	$\overline{x}_2$	$x_2 \rightarrow x_1$
x_1	x_2	f_8	f_9	f_{10}	f_{11}
0	0	1	1	1	1
0	1	0	0	0	0
1	0	0	0	1	1
1	1	0	1	0	1

Notation		$\overline{x}_1$	$x_1 \rightarrow x_2$	$x_1 \mid x_2$	1
x_1	x_2	f_{12}	f_{13}	f_{14}	f_{15}
0	0	1	1	1	1
0	1	1	1	1	1
1	0	0	0	1	1
1	1	0	1	0	1

The set of Boolean functions is called **universal for classical calculations**, if an arbitrary Boolean function can be expressed using functions from this set. Here are some examples of universal sets.

(1) The set $\{^{-}, \vee, \wedge\}$, consisting of negation, conjunction, and disjunction, is universal in the sense of this definition.

(2) **Reed[2]–Muller[3] basis**, or **Zhegalkin[4] basis** $\{\oplus, \wedge, 1\}$, also satisfies the introduced definition and therefore is universal in terms of classical calculations. In particular, the Boolean functions $g(x_1, x_2, x_3) = (x_1 \wedge x_2) \oplus (x_1 \downarrow x_3)$ and

---

[2]Irving Stoy Reed (1923–2012).
[3]David Eugene Muller (1924–2008).
[4]Ivan Ivanovich Zhegalkin (1869–1947).

Table 6.4 Three qubits operations

Name	Notation	Matrix
Toffoli gate		See formula (6.6)
Fredkin gate		See formula (6.8)

$h(x_1, x_2, x_3, x_4) = \overline{(x_1 \rightarrow x_2)} \rightarrow (x_3 \leftrightarrow x_4)$ can be represented as follows:

$$g(x_1, x_2, x_3) = 1 \oplus x_1 \oplus x_3 \oplus x_1 x_2 \oplus x_1 x_3, \tag{6.4}$$

$$h(x_1, x_2, x_3, x_4) = 1 \oplus x_1 x_3 \oplus x_1 x_4 \oplus x_1 x_2 x_3 \oplus x_1 x_2 x_4. \tag{6.5}$$

(3) A single-element set $\{|\}$ is universal, where "$|$" is Sheffer[5] stroke, operating in accordance with the rule $x_1 | x_2 = \overline{x_1 \wedge x_2}$. As is known, this operation corresponds to the element NAND in the theory of classical logical circuits. Let us note that with the combination of only NAND elements, it is possible to implement the computation of an arbitrary Boolean function if it is possible to copy the bit [3].

But, for example, the set $\{\oplus, \vee, \wedge\}$ is not universal, because the constant 1 cannot be expressed using only conjunction, disjunction, and addition modulo two.

In this section, we introduce two more quantum logic gates with the following property—each of them is universal for classical computations. Note that as an additional condition, we assume the possibility to include auxiliary inputs whose initial states are set to $|0\rangle$ or $|1\rangle$ as needed.

Table 6.4 represents universal quantum gates that transform the state of three qubits.

The **Toffoli**[6] **gate** is a quantum gate with three inputs and outputs that takes the input states $|\psi_1\rangle$, $|\psi_2\rangle$, and $|\psi_3\rangle$ and inverts the state $|\psi_3\rangle$ if and only if both first states are equal to $|1\rangle$ (and does nothing otherwise) [4].

Let us prove the universality of the Toffoli gate. Following from the definition, the first and second qubits on the input and output do not change. The third qubit in the output has the form $|\psi_3\rangle = |\psi_3 \oplus (\psi_1 \wedge \psi_2)\rangle$. Let two states $|\psi_1\rangle, |\psi_2\rangle$ and auxiliary qubit $|1\rangle$ be submitted to the gate input, respectively. So in the output, we will have the states $|\psi_1\rangle, |\psi_2\rangle$, and $|1 \oplus (\psi_1 \wedge \psi_2)\rangle$. The last of these states, as can be easily seen, can be represented as $|\overline{(\psi_1 \wedge \psi_2)}\rangle$, or $|(\psi_1 \text{ NAND } \psi_2)\rangle$. You can see that in this case, the Sheffer stroke is implemented, which as shown above on page 30, forms a universal set of functions. Consequently, the Toffoli gate is universal.

---

[5]Henry Maurice Sheffer (1882–1964).
[6]Tommaso Toffoli (born 1943).

The Toffoli gate matrix looks as follows:

$$U_{CCNOT} = \begin{bmatrix} 1 & 0 & 0 & 0 & 0 & 0 & 0 & 0 \\ 0 & 1 & 0 & 0 & 0 & 0 & 0 & 0 \\ 0 & 0 & 1 & 0 & 0 & 0 & 0 & 0 \\ 0 & 0 & 0 & 1 & 0 & 0 & 0 & 0 \\ 0 & 0 & 0 & 0 & 1 & 0 & 0 & 0 \\ 0 & 0 & 0 & 0 & 0 & 1 & 0 & 0 \\ 0 & 0 & 0 & 0 & 0 & 0 & 0 & 1 \\ 0 & 0 & 0 & 0 & 0 & 0 & 1 & 0 \end{bmatrix}. \tag{6.6}$$

By combining different input states, we can obtain the following functions:

$|\psi_1\rangle = |1\rangle$, $|\psi_2\rangle = |1\rangle$, $|\psi_3 \oplus (\psi_1 \wedge \psi_2)\rangle = \sigma_1 |\psi_3\rangle$, negation of $|\psi_3\rangle$,

$|\psi_3\rangle = |0\rangle$, $|\psi_3 \oplus (\psi_1 \wedge \psi_2)\rangle = |\psi_1 \wedge \psi_2\rangle$, conjunction,

$|\psi_2\rangle = |1\rangle$, $U_{CCNOT} |\psi_3\rangle = |\psi_1 \oplus \psi_3\rangle$, addition modulo two, XOR. (6.7)

We can see that the gate executes the operations of negation, disjunction, and addition modulo two depending on different values of states on the inputs.

With the help of Toffoli gate, two CNOTs, and one auxiliary qubit in the state $|0\rangle$, it is possible to implement the disjunction:

The **Fredkin[7] gate** is designated as Controlled SWAP (CSWAP) [5]. This gate has three inputs and three outputs. The first input is the control input, the other two are controlled inputs:

- if the control qubit equals $|0\rangle$, the controlled qubits do not change;
- if the control qubit is set to $|1\rangle$, then the values of the controlled qubits are exchanged (SWAP operation in relation to the controlled qubits).

The control line signal remains unchanged in any case. Let us write a matrix representation of the Fredkin gate:

$$U_{CSWAP} = \begin{bmatrix} 1 & 0 & 0 & 0 & 0 & 0 & 0 & 0 \\ 0 & 1 & 0 & 0 & 0 & 0 & 0 & 0 \\ 0 & 0 & 1 & 0 & 0 & 0 & 0 & 0 \\ 0 & 0 & 0 & 1 & 0 & 0 & 0 & 0 \\ 0 & 0 & 0 & 0 & 1 & 0 & 0 & 0 \\ 0 & 0 & 0 & 0 & 0 & 0 & 1 & 0 \\ 0 & 0 & 0 & 0 & 0 & 1 & 0 & 0 \\ 0 & 0 & 0 & 0 & 0 & 0 & 0 & 1 \end{bmatrix}. \tag{6.8}$$

[7]Edward Fredkin (born 1934).

Fig. 6.1 Controlled unitary operation U

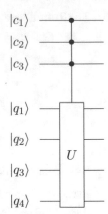

It is often useful to use an analytic view to describe the action of the Fredkin gate on the initial state $|\psi_{in}\rangle = |c, q_1, q_2\rangle$:

$$U_{CSWAP} |cq_1q_2\rangle = |c, \bar{c}q_1 \oplus cq_2, cq_1 \oplus \bar{c}q_2\rangle. \qquad (6.9)$$

One of the properties of the CSWAP element is also that the number of input states equal to $|1\rangle$ does not change as a result of its application.

The Fredkin gate is universal, as can be seen from the following particular cases of output values:

$$q_1 = 0, q_2 = 1: \quad U_{CSWAP} |c01\rangle = |c, c, \bar{c}\rangle, q_2 \to \bar{c}; \qquad (6.10)$$

$$q_2 = 0: \quad U_{CSWAP} |cq_10\rangle = |c, \bar{c}q_1, cq_1\rangle, q_2 \to c \wedge q_1; \qquad (6.11)$$

$$q_2 = 1: \quad U_{CSWAP} |cq_11\rangle = |c, \bar{c}q_1 \oplus c, cq_1 \oplus \bar{c}\rangle, q_1 \to c \vee q_1. \qquad (6.12)$$

Relations (6.10)–(6.11) show that implementations of operations such as negation, conjunction, and disjunction are available, i.e. a full set of Boolean functions is formed.

For many quantum algorithms, gates of controlled unitary operations are used, where some unitary transformation U over several qubits is performed or is not performed depending on the state of the control qubit set. The schematic representation of such gates is similar to that of a Toffoli gate (see Fig. 6.1).

For example, Fig. 6.1 shows a 4-qubit unitary operation gate U, controlled by three qubits $|c_1\rangle$, $|c_2\rangle$, and $|c_3\rangle$.

It is not difficult to construct the matrix representation of gates, controlled by a set of qubits. For example, in the matrix representation, the gate of some two-qubit unitary operation U controlled by N qubits has the form:

$$U_{NContrU} = \left.\begin{bmatrix} 1 & 0 & 0 & \cdots & 0 & 0 & 0 \\ 0 & 1 & 0 & \cdots & 0 & 0 & 0 \\ 0 & 0 & 1 & \cdots & 0 & 0 & 0 \\ \cdots & \cdots & \cdots & \cdots & \cdots & \cdots & \cdots \\ 0 & 0 & 0 & \cdots & 1 & 0 & 0 \\ 0 & 0 & 0 & \cdots & 0 & U_{11} & U_{12} \\ 0 & 0 & 0 & \cdots & 0 & U_{21} & U_{22} \end{bmatrix}\right\} N \text{ lines.} \qquad (6.13)$$

Table 6.5 Qubits states

Step 0	Step 1	Step 2	Step 3
q_1	q_1	q_1	q_1
q_2	$\overline{q}_1 q_2$	$\overline{q}_1 q_2$	q_2
q_3	q_3	$(q_1 q_2)q_3 + \overline{(q_1 q_2)}q_4$	$\overline{(q_1 q_2)}q_3 + (q_1 q_2)q_4$
q_4	q_4	$(q_1 q_2)q_3 + \overline{(q_1 q_2)}q_4$	$(q_1 q_2)q_3 + \overline{(q_1 q_2)}q_4$
0	$q_1 q_2$	$q_1 q_2$	0

In (6.13), values U_{ij} for $i, j \in \{1, 2\}$ are elements of the matrix of the operator U.

Example 6.2 Let us show that the quantum circuit having four qubits $|q_1\rangle$–$|q_4\rangle$ at its input, and one auxiliary qubit $|0\rangle$:

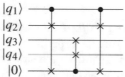

performs a SWAP operation on the third and fourth qubits controlled by the first and second qubits. In other words, such a circuit can be represented as

Solution.

Let us make a table of qubit states in the process of time evolution.

From Table 6.5, you can see that if the first and second qubits are simultaneously in the state $|1\rangle$, the states of the third and fourth qubits are exchanged. Otherwise, all states remain the same.   □

Example 6.3 Using Toffoli gates and one auxiliary qubit $|0\rangle$, let us build a quantum circuit for the NOT operation controlled by three qubits.

Solution.

Let us denote auxiliary qubits by c_1, c_2, and c_3. In order to build a quantum operation NOT controlled by these qubits, first calculate the conjunction $c_1 \wedge c_2$. This can be done with the help of one Toffoli gate. The result of the calculation will be written down using an auxiliary qubit. Next, using another Toffoli gate with control qubits $|c_1 c_2\rangle$, $|c_3\rangle$ and controlled qubit $|q\rangle$ will lead to the desired state $|q \oplus c_1 c_2 c_3\rangle$. Finally, the last gate returns the auxiliary qubit to its initial state.

Represent the proposed quantum circuit:

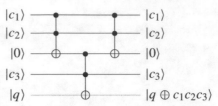

Note that the proposed procedure for building the NOT operation allows a generalization in case of an arbitrary finite number n of control qubits. In fact, the quantum circuit for operation $\underbrace{CC\ldots C}_{n \text{ times}} NOT$ is built from two symmetrically arranged circuits

for $\underbrace{CC\ldots C}_{n-1 \text{ times}} NOT$ and one Toffoli gate placed in the center. Exercise 49 shows that

such a circuit will contain an exponential number of elementary gates. Despite this, by applying more auxiliary qubits, it is easy to achieve a linear complexity of the circuit for operation $\underbrace{CC\ldots C}_{n-1 \text{ times}} NOT$ [4].   □

The implementation of unitary transformations with many qubits brings significant challenges to researchers from an experimental point of view. Therefore, it is of interest to make quantum circuits from the simplest, for example, single- and two-qubit elements. Let us show that the universal Toffoli and Fredkin gates are easy to express through the composition of simpler gates.

The following two circuits use a unitary transformation Q with a matrix:

$$U_Q = \frac{1}{2}(1 + i)\begin{bmatrix} 1 & -i \\ -i & 1 \end{bmatrix}. \tag{6.14}$$

This property is called the "square root of inversion", $\sqrt{NOT}$. It is easy to check the unitarity of the introduced operator:

$$U_Q U_Q^{\dagger} = \frac{1}{2}(1 + i)\begin{bmatrix} 1 & -i \\ -i & 1 \end{bmatrix} \times \frac{1}{2}(1 - i)\begin{bmatrix} 1 & i \\ i & 1 \end{bmatrix}$$

$$= \frac{1}{4}(1 + i)(1 - i)\begin{bmatrix} 2 & 0 \\ 0 & 2 \end{bmatrix} = \begin{bmatrix} 1 & 0 \\ 0 & 1 \end{bmatrix}. \tag{6.15}$$

Implementation of the Toffoli gate

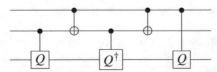

To prove that the above quantum circuit implements the Toffoli gate, let us note the following.

(1) If the first qubit is in the state $|0\rangle$, then as a result, the identity operator I or product of operators $QQ^{\dagger}$ affects the third qubit. Due to the unitarity of $QQ^{\dagger} \equiv I$, the third qubit does not change.

(2) If the first qubit is in the state $|1\rangle$, and the second qubit is in the state $|0\rangle$, then, analogous to the first case, $Q^\dagger Q$ affects the third qubit. The state of the third qubit is not affected.

(3) Finally, if the states of the upper qubits are $|1\rangle$, then the third qubit $|q\rangle$ transforms to $QQ|q\rangle \equiv U_{\text{NOT}}|q\rangle$.

The conditions (1)–(3) match the rules of the Toffoli gate action.

Implementation of the Fredkin gate

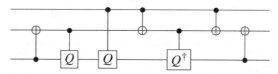

It should be noted that in the proposed implementation of the Fredkin gate, the Toffoli gate is taken as a basis, bordered in the right and left borders of the circuit by CNOT elements. It is proved that such a circuit is optimal in the number of used elementary gates [6].

Example 6.4 On the Quantum Computing Course exam, a student states that a Toffoli element can be implemented with just three controlled elements as follows:

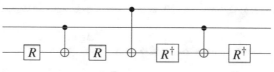

Here is used the designation $R = \begin{bmatrix} \cos(\pi/8) & \sin(\pi/8) \\ -\sin(\pi/8) & \cos(\pi/8) \end{bmatrix}$.

Is the student right?

Solution.

No, the student's statement is false. The state $|10\rangle\,|0\rangle$ will go to $|10\rangle\,(-|0\rangle)$ as a result of the work of this scheme, i.e. the controlled qubit will change the phase, while the Toffoli gate will leave the set of qubits $|100\rangle$ unchanged.

Let us prove it. Suppose that control qubits form the state $|q_1 q_2\rangle = |10\rangle$, and let us calculate how a controlled qubit $|q_3\rangle = u\,|0\rangle + v\,|1\rangle$, where $u, v \in \mathbb{C}$, is converted:

$$|q_3\rangle \rightarrow |q_3'\rangle = M_{R^\dagger} M_{R^\dagger} M_{\text{NOT}} M_R M_R \,|q_3\rangle. \tag{6.16}$$

Remember that the reverse order of matrices for constructing a product of operators is used (see page 11). With regard to

$$R^\dagger = \begin{bmatrix} \cos(\pi/8) & -\sin(\pi/8) \\ \sin(\pi/8) & \cos(\pi/8) \end{bmatrix}, \tag{6.17}$$

we perform matrix calculations:

$$\begin{aligned}
|q_3'\rangle = &\begin{bmatrix} \cos(\pi/8) & -\sin(\pi/8) \\ \sin(\pi/8) & \cos(\pi/8) \end{bmatrix} \begin{bmatrix} \cos(\pi/8) & -\sin(\pi/8) \\ \sin(\pi/8) & \cos(\pi/8) \end{bmatrix} \begin{bmatrix} 0 & 1 \\ 1 & 0 \end{bmatrix} \\
&\times \begin{bmatrix} \cos(\pi/8) & -\sin(\pi/8) \\ \sin(\pi/8) & \cos(\pi/8) \end{bmatrix} \begin{bmatrix} \cos(\pi/8) & -\sin(\pi/8) \\ \sin(\pi/8) & \cos(\pi/8) \end{bmatrix} \begin{bmatrix} u \\ v \end{bmatrix}.
\end{aligned} \tag{6.18}$$

After simple calculations, we come to the answer:

$$|q_3'\rangle = \begin{bmatrix} -u \\ v \end{bmatrix} \neq |q_3\rangle , \tag{6.19}$$

but, as it is known, the Toffoli gate does not change the third qubit in the state $|100\rangle$.

Note that the analyzed circuit provides a way to build a Toffoli gate *accurate to a phase factor* [7].   □

References

1. Rosen KH (ed) (2018) Handbook of discrete and combinatorial mathematics. Discrete mathematics and its applications, 2nd edn. CRC Press, xxiv, 1590 p
2. Rosen KH (2019) Discrete mathematics and its applications, 8th edn. McGraw-Hill, New York, NY, xxi, 942 p
3. Kurgalin S, Borzunov S (2020) The discrete math workbook: A companion manual using Python, 2nd edn. (Series: Texts in computer science). Springer, Cham, xvii, 500 p
4. Toffoli T (1980) Reversible computing. In: de Bakker JW, van Leeuwen J (eds) Proceedings of the 7th colloquium on automata, languages and programming. Springer, pp 632–644
5. Fredkin E, Toffoli T (1982) Conservative logic. Int J Theor Phys 21(3):219–253
6. Yu N, Ying M (2015) Optimal simulation of Deutsch gates and the Fredkin gate. Phys Rev A 91: 032302
7. Barenco A, Bennett CH, Cleve R, DiVincenzo DP, Margolus N et al (1995) Elementary gates for quantum computation. Phys Rev A 52(5):3457–3467

Implementation of Boolean Functions

7

The main task of classical computations is the computation of functions defined in analytical or tabular form. Let us start with Boolean functions that depend on several arguments.

The implementation of Boolean functions with a quantum computer is based on the construction of a quantum circuit that depends on the kind $f(\mathbf{x})$.

We represent the given function using only operations of conjunctions and addition modulo two. This representation is called **algebraic normal form** (ANF), or **Reed–Muller expansion**, or **Zhegalkin polynomial**.

The **algebraic normal form** of the function $f(\mathbf{x})$ is the sum modulo two of several elementary conjunctions of the form

$$G = K_1 \oplus K_2 \oplus \ldots \oplus K_s, \tag{7.1}$$

where K_i, $i = 1, 2, \ldots, s$ are pairwise different monotone elementary conjunctions over some set of variables $\{x_1, \ldots, x_n\}$, where $n = 1, 2, \ldots$. One of the K_i can be a constant one. The greatest of ranks of elementary conjunctions included in the polynomial G is called a **degree** of the function. It is known that any Boolean function is uniquely represented as a Reed–Muller decomposition accurate to the order of summands in the sum and the order of cofactors in the conjunctions [1].

Note. More strictly, the decomposition (7.1) is called **positive polarity Reed–Muller expansion**.

There are several methods of constructing an algebraic normal form $G(\mathbf{x})$, expressing the given function $f(x_1, x_2, \ldots, x_n)$. Next, let us consider two of these methods in detail.

1. *Method of undetermined coefficients.*

The Zhegalkin polynomial has the following form:

$$G(x_1, x_2, \ldots, x_n) = g_0 \wedge 1 \oplus g_1 \wedge K_1 \oplus g_2 \wedge K_2 \oplus \ldots \oplus g_{2^n-1} \wedge K_{2^n-1}, \tag{7.2}$$

where K_i are monotone elementary conjunctions, and coefficients $g_i \in \mathbb{B}$, $i = 0, 1, \ldots, 2^n - 1$.

© The Author(s), under exclusive license to Springer Nature Switzerland AG 2021
S. Kurgalin and S. Borzunov, *Concise Guide to Quantum Computing*,
Texts in Computer Science, https://doi.org/10.1007/978-3-030-65052-0_7

To determine the unknown coefficients $g_0, g_1, \ldots, g_{2^n-1}$, we make equations $G(\mathbf{x}_j) = f(\mathbf{x}_j)$ for all sorts of tuples $\mathbf{x}_j$. We get a system of 2^n equations with 2^n unknowns; its solution will give coefficients of the polynomial $G(\mathbf{x})$.

2. *Method of equivalent transformations.*

Let us write down an analytic expression for the function $f(\mathbf{x})$ and reduce it with identity transformations using the equality $A \vee B = \overline{(\overline{A} \wedge \overline{B})}$ to an equivalent expression containing only operations from the set $\{^-, \wedge\}$. Next, we exclude negation operations, for which we will replace everywhere expressions of the form $\overline{A}$ with $A \oplus 1$. Then we expand the brackets using the distribution law $A \wedge (B \oplus C) = (A \wedge B) \oplus (A \wedge C)$ and taking into account the equivalencies

$$A \wedge A = A, \quad A \wedge 1 = A,$$
$$A \oplus A = 0, \quad A \oplus 0 = A.$$

As a result, we get algebraic normal form for the function $f(\mathbf{x})$.

Note that if the function is given by a vector of values, in some cases it is convenient to switch to its disjunctive normal form, and only then to calculate the Reed–Muller expansion using, for example, the method of equivalent transformation.

Example 7.1 Let us build an algebraic normal form for the function $f(x_1, x_2, x_3)$, defined by the vector of values $\alpha = (1010\,0100)$.

Solution.

We use the method of undetermined coefficients. The algebraic normal form for the function $f(x_1, x_2, x_3)$ has the following form:

$$G(x_1, x_2, x_3) = g_0 \oplus g_1 x_1 \oplus g_2 x_2 \oplus g_3 x_3$$
$$\oplus g_4 x_1 x_2 \oplus g_5 x_1 x_3 \oplus g_6 x_2 x_3 \oplus g_7 x_1 x_2 x_3, \qquad (7.3)$$

where $g_0, \ldots, g_7 \in \mathbb{B}$.

Using the conditions $f(0,0,0) = 1$, $f(0,0,1) = 1$, $f(0,1,0) = 1$, etc. we will get a system of equations with respect to unknown coefficients $g_0, \ldots, g_7$:

$$\begin{cases}
g_0 = 1, \\
g_0 \oplus g_3 = 0, \\
g_0 \oplus g_2 = 1, \\
g_0 \oplus g_2 \oplus g_3 \oplus g_6 = 0, \\
g_0 \oplus g_1 = 0, \\
g_0 \oplus g_1 \oplus g_3 \oplus g_5 = 1, \\
g_0 \oplus g_1 \oplus g_2 \oplus g_4 = 0, \\
g_0 \oplus g_1 \oplus g_2 \oplus g_3 \oplus g_4 \oplus g_5 \oplus g_6 \oplus g_7 = 0.
\end{cases} \qquad (7.4)$$

The solution to this system are Boolean values $g_0 = 1$, $g_1 = 0$, $g_2 = 0$, $g_3 = 1$, and $g_4 = g_5 = g_6 = g_7 = 0$. Finally, let us write the algebraic normal form for the function $f(x_1, x_2, x_3)$:

$$f(x_1, x_2, x_3) = 1 \oplus x_1 \oplus x_3 \oplus x_1 x_2 x_3. \qquad (7.5)$$

On the basis of the algebraic normal form, construct a quantum circuit implementing an arbitrary Boolean function $f(x_1, x_2, \ldots, x_n)$ [2,3]. For this purpose, we use an n-qubit register reflecting the input data and a separate qubit $|q\rangle$ for the output of the response. The state of the quantum system is described by the product $|x_0, x_1, \ldots, x_n\rangle |q\rangle$. Qubit $|q\rangle$ takes the initial value $|0\rangle$.

The method of constructing the quantum circuit is as follows [4].

For each of the Reed–Muller expansion summands (7.1), add a NOT gate controlled by qubits with variables from these summands.

- The constant $f = 1$ is represented by a standard NOT gate.
- Variables of the type x_i, $1 \leqslant i \leqslant n$, are represented by the inversion NOT controlled by the qubit $|x_i\rangle$.
- Paired conjunctions of the form $x_i x_j$, $1 \leqslant i, j \leqslant n$, are represented by the inversion controlled by two qubits $|x_i\rangle$ and $|x_j\rangle$.
- Triple conjunctions of the form $x_i x_j x_k$, $1 \leqslant i, j, k \leqslant n$, are represented by the inversion controlled by three qubits $|x_i\rangle$, $|x_j\rangle$, and $|x_k\rangle$.

As it is easy to see, using the NOT, CNOT, CCNOT, ..., CC...CNOT operations, it is possible to implement an arbitrary Boolean function. The order of applying elementary operations does not matter, because they commutate with each other. This property is derived from the commutability of the addition modulo two operation in Boolean algebra:

$$\forall x_1, x_2 \in \mathbb{B} \ (x_1 \oplus x_2 = x_2 \oplus x_1). \tag{7.6}$$

Example 7.2 The Boolean function of three variables is defined as $f(x_1, x_2, x_3) = x_1 \vee x_1 \overline{x}_2 \vee \overline{x}_1 x_2 \overline{x}_3$. Build a quantum circuit that implements this function.

Solution.

First of all, we calculate the algebraic normal form of the function $f(x_1, x_2, x_3)$:

$$G_f(x_1, x_2, x_3) = x_1 \oplus x_2 \oplus x_1 x_2 \oplus x_2 x_3 \oplus x_1 x_2 x_3. \tag{7.7}$$

Thus, to implement $f(x_1, x_2, x_3)$, two CNOT elements, one CCNOT element, and one CCCNOT element are required. Their order in the circuit is arbitrary; for certainty, we choose the order defined by the formula (7.7).

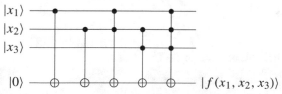

The obtained quantum circuit implements the Boolean function $f(x_1, x_2, x_3) = x_1 \vee x_1 \overline{x}_2 \vee \overline{x}_1 x_2 \overline{x}_3$. □

If necessary, an arbitrary circuit containing controlled inversions can be brought to a form containing only two-qubit elements (see page 33).

As in the case of classical computations, it is possible to implement more complex types of functional dependencies using Boolean functions. Let us show how to calculate the function of an integer non-negative argument.

$$f\colon \mathbb{Z}_0 \to \mathbb{Z}_0, \tag{7.8}$$

where $\mathbb{Z}_0 = \{0, 1, 2, \dots\}$.

To solve this problem, we code the argument in an n-qubit register, and the result will be read from another, m-qubit register.

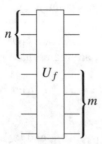

In this way it is possible to input the numbers $0, 1, 2, \dots, 2^n - 1$, and to obtain the result from the set $[0, 2^m - 1]$. If it is necessary to extend either the domain or the range of the function values, the n or m parameter should be increased, respectively.

Example 7.3 Let us construct a quantum circuit implementing the function $f(x) = x + 1$. Consider that the domain of the function is the set $\{0, 1, 2, 3\}$.

Solution.

We make a table of values of the function $f(x)$, where we specify the binary representation of the argument $(x)_{\text{bin}}$.

From Table 7.1, we can see that two qubits $|x_1\rangle$ and $|x_2\rangle$ are sufficient to encode the argument, and three qubits $|q_1\rangle$, $|q_2\rangle$, and $|q_3\rangle$ are sufficient to encode the response.

Thus, the high bit of the response $[f(x)]_{\text{bin}}$ in the binary form has a vector value $\alpha = (0001)$, the second bit has a vector value $\beta = (0110)$, and the third bit has a vector value $\gamma = (1010)$. Therefore, the high bit implements the Boolean function $x_1 \wedge x_2$, the second bit corresponds to the Boolean function $x_1 \oplus x_2$, finally, the third bit corresponds to the Boolean function $1 \oplus x_2$ (see Figs. 7.1 and 7.2). So we get the following circuit:

Table 7.1 The table of values of $f(x)$

x	$(x)_{\text{bin}}$	$f(x)$	$[f(x)]_{\text{bin}}$
0	0 0	1	0 0 1
1	0 1	2	0 1 0
2	1 0	3	0 1 1
3	1 1	4	1 0 0
qubits	$x_1 x_2$		$q_1 q_2 q_3$

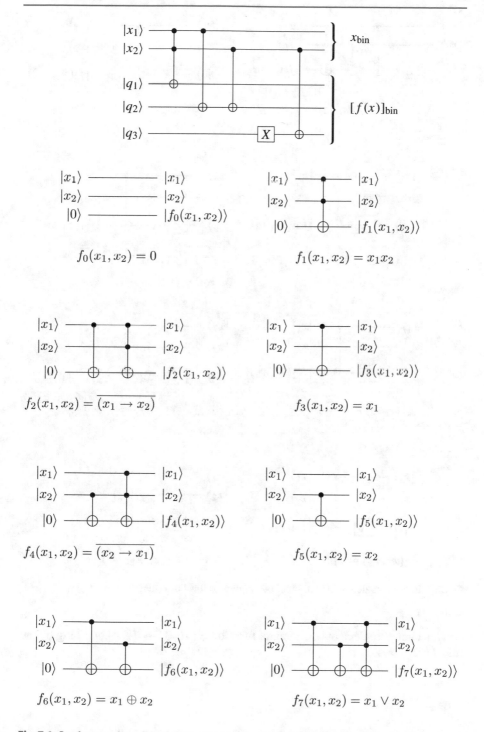

Fig. 7.1 Implementation of the full set of two-argument functions (functions f_0–f_7)

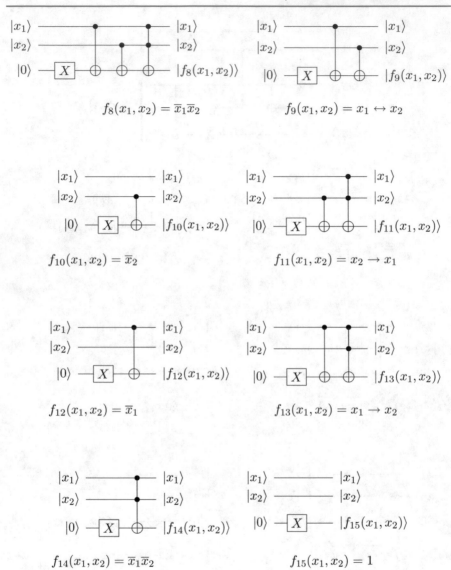

Fig. 7.2 Implementation of the full set of two-argument functions (functions f_8–f_{15})

This circuit implements the integral function $f : [0, 3] \to [1, 4]$, working by the rule $f(x) = x + 1$.   □

References

1. Koshy T (2004) Discrete mathematics with applications. Elsevier Academic Press, xxiv, 1045 p
2. Ngoc VP, Wiklicky H (2020) Tunable quantum neural networks for Boolean functions. ArXiv:2003.14122v1 [quant-ph]
3. Younes A, Miller J (2018) Automated method for building CNOT based quantum circuits for Boolean functions. ArXiv:quant-ph/0304099
4. Younes A, Miller J (2004) Representation of Boolean quantum circuits as Reed–Muller expansions. Int J Electron 91(7):431–444

Quantum Fourier Transform

8

Let us recall the main definitions associated with the classical Fourier[1] transform known from the course of mathematical analysis.

The Fourier transform is one of the central methods of modern applied mathematics [1,2].

Consider a function $s : \mathbb{R} \to \mathbb{C}$. Let $s(t)$ be absolutely integrable on its domain of definition, that is $\int_{-\infty}^{\infty} |s(t)| \, dt < \infty$.

The **Continuous Fourier Transform** of the function $s(t)$ is determined by the relation

$$S(f) = \int_{-\infty}^{\infty} s(t) e^{2\pi i f t} \, dt. \tag{8.1}$$

Function $S(f)$ is called an **image** of the original function $s(t)$.

Note. Traditionally, $s(t)$ in physical applications is considered as a **signal**, and $S(f)$ is considered as a **signal spectrum**. Furthermore, the variable t has the meaning of time and f has the meaning of linear frequency. Note that the designation $S(f) = \mathcal{F}[s(t)]$ is widely used.

In computational problems, the considered function is usually defined not analytically, but in the form of discrete values on some grid, usually uniform.

In practical problems, the sample length

$$\mathbf{x} = (\ldots, x_{-2}, x_{-1}, x_0, x_1, x_2, \ldots) \tag{8.2}$$

is usually bounded, so let us consider the transition from a continuous integral transform $\mathcal{F}[s(t)]$ to an approximate integral sum:

---

[1] Jean-Baptiste Joseph Fourier (1786–1830).

S. Kurgalin and S. Borzunov, *Concise Guide to Quantum Computing*, Texts in Computer Science, https://doi.org/10.1007/978-3-030-65052-0_8

$$(\mathcal{F}[s(t)])_n = \int_{-\infty}^{\infty} s(t) e^{2\pi i f_n t} dt \;\rightarrow\; \sum_{k=-\infty}^{\infty} x_k e^{2\pi i f_n t_k} \delta = \delta \sum_{k=0}^{N-1} x_k e^{2\pi i k n/N}. \quad (8.3)$$

Thus, we come to the definition of the **Discrete Fourier Transform** (DFT) of N values x_k, where $k = 0, 1, \ldots, N - 1$:

$$y_n = \sum_{k=0}^{N-1} x_k e^{2\pi i k n/N}, 0 \leqslant n \leqslant N - 1, \quad (8.4)$$

or, using the designation for the **root of unity** $\omega = e^{2\pi i/N}$,

$$y_n = \sum_{k=0}^{N-1} \omega^{kn} x_k, 0 \leqslant n \leqslant N - 1. \quad (8.5)$$

Note that the values ω and their integer non-negative degrees $\omega_k \equiv \omega^k = e^{2\pi i k/N}$ have a number of interesting properties, some of which have been investigated in Exercises 74–76.

The **Inverse Discrete Fourier Transform** is determined by the formula

$$x_n = \frac{1}{N} \sum_{k=0}^{N-1} \omega^{-kn} y_k, 0 \leqslant n \leqslant N - 1. \quad (8.6)$$

In a vector notation, the considered relations have the form $\mathbf{y} = \mathcal{F}[\mathbf{x}]$ and $\mathbf{x} = \mathcal{F}^{-1}[\mathbf{y}]$.

The sequential application of direct and inverse DFT to an arbitrary vector $\mathbf{x}$ does not change its components: $\forall \mathbf{x} \in \mathbb{R}^n \; \mathcal{F}^{-1}[\mathcal{F}[\mathbf{x}]] = \mathbf{x}$ and $\mathcal{F}[\mathcal{F}^{-1}[\mathbf{x}]] = \mathbf{x}$. Prove, for example, the first of these equations.

$$(\mathcal{F}^{-1}[\mathcal{F}[\mathbf{x}]])_n = \frac{1}{N} \sum_{k=0}^{N-1} \omega^{-kn} (\mathcal{F}[\mathbf{x}])_k = \frac{1}{N} \sum_{k=0}^{N-1} \omega^{-kn} \sum_{i=0}^{N-1} \omega^{ik} x_i. \quad (8.7)$$

Let us change the summation order:

$$(\mathcal{F}^{-1}[\mathcal{F}[\mathbf{x}]])_n = \sum_{i=0}^{N-1} x_i \left(\frac{1}{N} \sum_{k=0}^{N-1} \omega^{k(i-n)} \right). \quad (8.8)$$

Further, use the properties of the values $\omega = e^{2\pi i/N}$, considered in Exercise 75 (see also [3]):

$$\frac{1}{N} \sum_{k=0}^{N-1} \omega^{k(i-n)} = \begin{cases} 1, & \text{or } i = n; \\ 0, & \text{or } i \neq n. \end{cases} \quad (8.9)$$

Finally, we obtain $(\mathcal{F}^{-1}[\mathcal{F}[\mathbf{x}]])_n = x_n \; \forall n = 0, 1, \ldots, N - 1$. The equality $\mathcal{F}[\mathcal{F}^{-1}[\mathbf{x}]] = \mathbf{x}$ is proved in the same way.

According to the definition, the computation of the vector **y** for the given **x** requires $O(N^2)$ complex multiplications.[2] There is a way to significantly reduce the asymptotic complexity of the DFT. The **Fast Fourier Transform** algorithm, or **FFT**, requires only $O(N \log_2 N)$ multiplication operations; there are also ways to parallelize the FFT [6]. In further consideration, we will assume that $N = 2^m$ for some natural number m. This limitation is not fundamental, because one of the following approaches can be used if you need to implement the Fast Fourier Transform algorithm for $N \neq 2^m$:

(1) fill in several cells of the array representing a signal with zeros, so that N becomes equal to the nearest power of two;

(2) use more complex generalizations of the FFT (see, for example, [7]).

To distribute FFT operations among computational nodes, the following approach [8] is used: the vector $\mathbf{x} = (x_0, \ldots, x_{N-1})$, representing input signal counts, is considered as a two-dimensional array of size $N_1 \times N_2$, where $N_1 = 2^{m_1}$, $N_2 = 2^{m_2}$ for some $m_1, m_2 \in \mathbb{N}$, such that $2^{m_1+m_2} = N$. Then the index i of an arbitrary element of vector **x** can be written in the form

$$i = LN_1 + l, \text{ where} 0 \leqslant l \leqslant N_1 - 1, 0 \leqslant L \leqslant N_2 - 1.$$

Components of Fourier transform $\mathbf{y} = \mathcal{F}[\mathbf{x}]$ are calculated according to the formula

$$y_n = \sum_{k=0}^{N_1 N_2 - 1} e^{\frac{2\pi i k n}{N_1 N_2}} x_k \tag{8.10}$$

for all $0 \leqslant n \leqslant N_1 N_2 - 1$, or

$$y_{\tilde{l} N_2 + \tilde{L}} = \sum_{L=0}^{N_2 - 1} \sum_{l=0}^{N_1 - 1} e^{\frac{2\pi i}{N_1 N_2}(\tilde{l} N_2 + \tilde{L})(L N_1 + l)} x_{L N_1 + l}. \tag{8.11}$$

After algebraic transformations, let us write the components of the vector **y** in the following form:

$$y_{\tilde{l} N_2 + \tilde{L}} = \sum_{l=0}^{N_1 - 1} e^{\frac{2\pi i \tilde{l} l}{N_1}} \left(e^{\frac{2\pi i l \tilde{L}}{N_1 N_2}} \left(\sum_{L=0}^{N_2 - 1} e^{\frac{2\pi i L \tilde{L}}{N_2}} x_{L N_1 + l} \right) \right). \tag{8.12}$$

---

[2] An asymptotic notation $O(g(n))$ is used here. $O(g(n))$ is a **class of functions growing no faster than** $g(n)$ [4,5]. It is defined as

$$O(g(n)) = \{f(n): \exists c \in (0, \infty), \ n_0 \in \mathbb{N} \text{ such that for all } n \geqslant n_0$$
$$\text{the inequalities } 0 \leqslant f(n) \leqslant cg(n) \text{ are valid}\}.$$

It is said that the function $f(n)$ belongs to the class $O(g(n))$ (read as "big-O of g"), if for all values of the argument n, starting from the threshold value $n = n_0$, the inequalities $f(n) \leqslant cg(n)$ are valid for some positive c. Also, a **class of functions growing at the same rate as** $g(n)$ is used, and it is denoted by $\Theta(g(n))$.

The implementation of a parallel program based on the obtained expression for $\mathbf{y}$ is easy. Using a possible parallelism over the variable l, there is calculated the inner sum

$$\sum_{L=0}^{N_2-1} e^{\frac{2\pi i L \tilde{L}}{N_2}} x_{LN_1+l} \tag{8.13}$$

for $l = 0, \ldots, N_1-1$. The values are then multiplied by $e^{\frac{2\pi i l \tilde{L}}{N_1 N_2}}$; then another FFT with the possibility of parallelism over the variable L is applied, where $L = 0, \ldots, N_2-1$. The values $y_{\tilde{l}N_2+\tilde{L}}$ form the final answer $\mathbf{y} = \mathcal{F}[\mathbf{x}]$.

The **Quantum Fourier Transform** is a component of many quantum algorithms, including the quantum algorithm of the decomposition of a compound number into prime multipliers, the algorithm of phase evaluation for finding unitary operator eigenvalues, etc.

Let us consider the orthonormalized computational basis

$$B = \{|0\rangle, |1\rangle, \ldots, |N-1\rangle\}, \tag{8.14}$$

where N is the number of states that form the basis. The **Quantum Discrete Fourier Transform** of computational basis elements is determined using the following formula:

$$|j\rangle \to \mathcal{F}[|j\rangle] = \frac{1}{\sqrt{N}} \sum_{k=0}^{N-1} e^{2\pi i j k/N} |k\rangle. \tag{8.15}$$

The action of the operator $\mathcal{F}$ on an arbitrary quantum state, represented as a superposition of basis states $|j\rangle$, is equal to

$$\sum_{j=0}^{N-1} c_j |j\rangle \to \mathcal{F}\left[\sum_{j=0}^{N-1} c_j |j\rangle\right] = \frac{1}{\sqrt{N}} \sum_{j,k=0}^{N-1} c_j e^{2\pi i j k/N} |k\rangle. \tag{8.16}$$

This formula reflects the quantum-mechanical principle of the superposition of states.

Matrix F_N of the operator $\mathcal{F}$ has the size of $N \times N$, and it is equal to

$$F_N = \frac{1}{\sqrt{N}} \begin{bmatrix} 1 & 1 & 1 & \cdots & 1 \\ 1 & \omega_n & \omega_n^2 & \cdots & \omega_n^{N-1} \\ 1 & \omega_n^2 & \omega_n^4 & \cdots & \omega_n^{2(N-1)} \\ 1 & \omega_n^3 & \omega_n^6 & \cdots & \omega_n^{3(N-1)} \\ \cdots for 5 & & & & \\ 1 & \omega_n^{N-1} & \omega_n^{2(N-1)} & \cdots & \omega_n^{(N-1)(N-1)} \end{bmatrix}. \tag{8.17}$$

The elements of this matrix are defined by the formula

$$(F_N)_{k_1,k_2} = e^{2\pi i (k_1-1)(k_2-1)/N} \tag{8.18}$$

for all $k_1, k_2 = 0, 1, \ldots, N-1$.

Suppose that the size of the computational basis is equal to a positive integer degree of two: $N = 2^n$ for some $n \in \mathbb{N}$. For a further explanation, we will need a binary representation of j.

We introduce a notation $j = j_1 j_2 \ldots j_n$ for the binary representation of the natural number j:

$$j = 2^{n-1} j_1 + 2^{n-2} j_2 + \cdots + 2^0 j_n. \tag{8.19}$$

Similarly, for real $0 < j < 1$ we write the binary representation as

$$j = (0.j_l j_{l+1} \ldots j_m) = 2^{-1} j_l + 2^{-2} j_{l+1} + \cdots + 2^{l-m-1} j_m. \tag{8.20}$$

In particular, $(0.j_1) = j_1/2$, $(0.j_1 j_2) = j_1/2 + j_2/4$, etc.

Taking into account the introduced notations, the Quantum Fourier Transform can be represented as follows:

$$|j_1 j_2 \ldots j_n\rangle \rightarrow \mathcal{F}[|j_1 j_2 \ldots j_n\rangle]$$
$$= \frac{1}{2^{n/2}} \left(|0\rangle + e^{2\pi i (0.j_n)} |1\rangle \right) \times \left(|0\rangle + e^{2\pi i (0.j_{n-1} j_n)} |1\rangle \right) \ldots$$
$$\times \left(|0\rangle + e^{2\pi i (0.j_1 j_2 \ldots j_n)} |1\rangle \right). \tag{8.21}$$

In fact, the following sequence of equations is executed:

$$\mathcal{F}_N[|j\rangle] = \frac{1}{\sqrt{N}} \sum_{k=0}^{N-1} e^{2\pi i j k 2^{-n}} |k\rangle$$

$$= \frac{1}{\sqrt{N}} \sum_{k_1 \in \{0,1\}} \sum_{k_2 \in \{0,1\}} \cdots \sum_{k_n \in \{0,1\}} e^{2\pi i j \left(\sum_{l=1}^{n} k_l 2^{-l} \right)} |k_1 k_2 \ldots k_n\rangle$$

$$= \frac{1}{\sqrt{N}} \sum_{k_1 \in \{0,1\}} \sum_{k_2 \in \{0,1\}} \cdots \sum_{k_n \in \{0,1\}} \prod_{l=1}^{n} e^{2\pi i j k_l 2^{-l}} |k_l\rangle$$

$$= \frac{1}{\sqrt{N}} \prod_{l=1}^{n} \sum_{k_l \in \{0,1\}} e^{2\pi i j k_l 2^{-l}} |k_l\rangle$$

$$= \frac{1}{\sqrt{N}} \prod_{l=1}^{n} \left(|0\rangle + e^{2\pi i j 2^{-l}} |1\rangle \right)$$

$$= \frac{1}{\sqrt{2}} \left(|0\rangle + e^{2\pi i (0.j_n)} |1\rangle \right) \times \frac{1}{\sqrt{2}} \left(|0\rangle + e^{2\pi i (0.j_{n-1} j!_n)} |1\rangle \right) \ldots$$

$$\times \frac{1}{\sqrt{2}} \left(|0\rangle + e^{2\pi i (0.j_1 j_2 \ldots j_{n-1} j_n)} |1\rangle \right). \tag{8.22}$$

Thus, the formula (8.21) is proven. Note that $\mathcal{F}_N[|k\rangle]$ is presented in a factorialized form, and the corresponding state is not confusing [9].

From the representation (8.21), if follows directly that the first output qubit is in the superposition $|0\rangle$ and $e^{2\pi i (0.j_1 \ldots j_n)} |1\rangle$, the second qubit is in the superposition $|0\rangle$ and $e^{2\pi i (0.j_2 \ldots j_n)} |1\rangle$, etc., and the final qubit in the superposition $|0\rangle$ and $e^{2\pi i (0.j_n)} |1\rangle$.

Let us turn to a number of special cases of small values n of the computational basis. Next, using the gate R_p, we will denote a unitary transformation of the form

$$R_p = \begin{bmatrix} 1 & 0 \\ 0 & e^{2\pi i/2^p} \end{bmatrix} \tag{8.23}$$

for $p = 2, 3, \ldots, n$.

1. *One qubit*: $n = 1$.

$$\mathcal{F}[|j_1\rangle] = \frac{1}{\sqrt{2}} \left(|0\rangle + e^{2\pi i (0.j_1)} |1\rangle \right) = \frac{1}{\sqrt{2}} \left(|0\rangle + e^{\pi i j_1} |1\rangle \right). \tag{8.24}$$

Since

$$e^{\pi i j_1} = \begin{cases} 1, & \text{if } j_1 = 0, \\ -1, & \text{if } j_1 = 1, \end{cases} \tag{8.25}$$

the resulting transformation matches the Hadamard transform (3.7):

$$|q_1\rangle \longrightarrow \boxed{H} \longrightarrow |y_1\rangle$$

2. *Two qubits*: $n = 2$.

$$|j_1 j_2\rangle \rightarrow \mathcal{F}[|j_1 j_2\rangle]$$
$$= \frac{1}{2} \left(|0\rangle + e^{2\pi i (0.j_2)} |1\rangle \right) \times \left(|0\rangle + e^{2\pi i (0.j_1 j_2)} |1\rangle \right). \tag{8.26}$$

The transformation is implemented by the following quantum circuit:

$$|q_1\rangle \longrightarrow \boxed{H} \longrightarrow \boxed{R_2} \longrightarrow \times \longrightarrow |y_1\rangle$$
$$|q_2\rangle \longrightarrow \bullet \longrightarrow \boxed{H} \longrightarrow \times \longrightarrow |y_2\rangle$$

3. *Three qubits*: $n = 3$.

$$|j_1 j_2 j_3\rangle \rightarrow \mathcal{F}[|j_1 j_2 j_3\rangle]$$
$$= \frac{1}{2^{n/2}} \left(|0\rangle + e^{2\pi i (0.j_3)} |1\rangle \right) \times \left(|0\rangle + e^{2\pi i (0.j_2 j_3)} |1\rangle \right)$$
$$\times \left(|0\rangle + e^{2\pi i (0.j_1 j_2 j_3)} |1\rangle \right). \tag{8.27}$$

Quantum circuit:

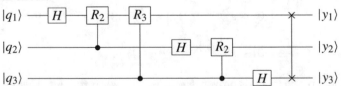

For a system formed by n qubits, we get the following circuit. Several SWAP exchanges on the final steps of the algorithm restore the required state order.

Let us note the important difference in the results of Classical and Quantum Fourier Transforms. The Fourier Transform (8.4), executed over a vector of values of the function $(x_1, x_2 \ldots, x_n)$, leads to a new vector $\mathbf{y} = (y_1, y_2, \ldots, y_n)$. All components of this vector are known. In the case of the Quantum Fourier Transform applied to the state of a set of qubits, a new state—a superposition of basis states—will be obtained. Information on Fourier coefficients is contained in the amplitudes of basis

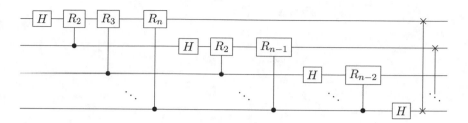

Fig. 8.1 Quantum Fourier Transform Circuit

states. A single measurement made at $|\psi_{out}\rangle = \mathcal{F}[|\psi_{in}\rangle]$ will not produce the full spectrum. Moreover, the fact of making a measurement will project $|\psi_{out}\rangle$ into one of the basis states. Quantum algorithms that use the Fourier Transform must take into account these limitations when working with the results of this transform (Fig. 8.1).

The Quantum Fourier Transform is implemented using a circuit containing only $O(n^2)$ elementary logic gates. This is much smaller than the transform (8.4), which requires $O(N \log N) = O(n \times 2^n)$ of elementary operations.

References

1. Grafakos L (2008) Classical fourier analysis, 2nd edn. (Series: Graduate texts in mathematics, vol 249). Springer, xvi, 489 p
2. Vretblad A (2003) Fourier analysis and its applications (Series: Graduate texts in mathematics, vol 223). Springer, xii, 269 p
3. Kurgalin S, Borzunov S (2020) The discrete math workbook: A companion manual using Python, 2nd edn. (Series: Texts in computer science). Springer, Cham, xvii, 500 p
4. Cormen TH, Leiserson CE, Rivest RL, Stein C (2009) Introduction to algorithms, 3rd edn. The MIT Press, Cambridge, Massachusetts, London, xx, 1292 p
5. McConnell JJ (2001) Analysis of algorithms: an active learning approach. Jones and Bartlett Publishers, Sudbury, Massachusetts
6. Kurgalin S, Borzunov S (2019) A practical approach to high-performance computing. Springer International Publishing, Cham, xi, 206 p
7. Van Loan C (1992) Computational frameworks for the fast Fourier transform (Series: Frontiers in applied mathematics). Society for Industrial and Applied Mathematics, Philadelphia, 273 p
8. Press WH et al (2007) Numerical recipes: the art of scientific computing, 3rd edn. Cambridge University Press, xxii, 1235 p
9. Billig Y (2018) Quantum computing for high school students, 134 p

Phase Evaluation Algorithm

9

In a number of quantum-mechanical problems, you need to define eigenvalues of some unitary operator U:

$$U |\psi\rangle = \lambda |\psi\rangle, \tag{9.1}$$

where $|\psi\rangle$ is a known eigenvector of operator U, λ is an eigenvalue, which needs to be defined. Due to the unitarity of U, the equality $|\lambda| = 1$ is fulfilled, and therefore the eigenvalue can be represented as $\lambda = e^{2\pi i \varphi}$, where φ is a real number (as it is said, the **transformation phase**) satisfies the condition $0 \leqslant \varphi < 1$.

Let the bit representation of the value φ is finite:

$$\varphi = 0.\varphi_1 \varphi_2 \ldots \varphi_n. \tag{9.2}$$

In the phase estimation algorithm, there will be used 2 quantum registers, the first of which contains n qubit, originally pre-set in $|0\rangle$. The second register is originally in the state $|\psi\rangle$, the number of qubits in this register depends on the specific type of the operator U. The algorithm consists of the following steps.

1. For $N = 2^n$, we compute the Direct Fourier Transform, applied to $\left| \underbrace{00\ldots0}_{n \text{ qubits}} \right\rangle |\psi\rangle$.

The result is equal to $\sum_{q=0}^{N-1} |q\rangle |\psi\rangle$.

2. Apply U to the second register q times:

$$|q\rangle |\psi\rangle \rightarrow |q\rangle U^q |\psi\rangle = e^{2\pi i \varphi q} |q\rangle |\psi\rangle. \tag{9.3}$$

The first n qubits form the state

$$\frac{1}{\sqrt{N}} \sum_{q=0}^{N-1} e^{2\pi i \varphi q} |q\rangle = \mathcal{F}[|2^n \varphi\rangle]. \tag{9.4}$$

3. Calculate the inverse Fourier Transform $\mathcal{F}^{-1}$, applied to the first n qubits and perform the measurement procedure. We obtain $|2^n \varphi\rangle = |\varphi_1 \varphi_2 \ldots \varphi_n\rangle$ with a probability, equal to one.

© The Author(s), under exclusive license to Springer Nature Switzerland AG 2021
S. Kurgalin and S. Borzunov, *Concise Guide to Quantum Computing*,
Texts in Computer Science, https://doi.org/10.1007/978-3-030-65052-0_9

It remains to consider the case when the bit representation of the value φ is not finite: $\varphi = 0.\varphi_1\varphi_2 \ldots \varphi_n \ldots$. Applying a phase evaluation algorithm leads to an approximate value of φ.

The phase estimation algorithm has widespread application in quantum computing, being a part of the algorithm of factorization of composite numbers and algorithm of solving systems of linear algebraic equations [1].

As can be seen from the examples of the reviewed quantum algorithms, quantum-mechanical computations have significant differences compared to the classical methods of computation organization. Thus, the result of quantum computing can be obtained only after measuring the system state, which is a non-deterministic (probabilistic) process. Due to this non-determined character is inherent to almost all quantum algorithms.

Reference

1. Harrow AW, Hassidim A, Lloyd S (2009) Quantum algorithm for linear systems of equations. Phys Rev Lett 103:150502

Quantum Search

The search for information in some data structure is an important task in the field of computer sciences. Traditionally, it is considered a simple data structure—an array with integer elements. Generalization of search algorithms into more complex structures, as a rule, does not cause any difficulties [1, 2].

Let us briefly remind you how the search problem is solved by classical methods.

Let there be a problem of determining the index of the array element equal in value to this particular value, which is called **target**. **Sequential search algorithm** is applied to work with an unordered array and operates in the following way: sequentially looks through the array elements starting from the first one and compares them with the target one. If the sought element is found, the index of this element is returned, otherwise the result must be a number that does not correspond to any element, for example, −1.

Analysis of the sequential search algorithm shows that when the target element is located at the end of an array, the number of required element operations is N, where N is the array size. For the complexity of this algorithm in the average case, we get the expression

$$A(N) = \frac{N + 1}{2} = \Theta(N), \tag{10.1}$$

which uses the asymptotic notation $\Theta(g(N))$ is the *class of functions growing at the same rate as* $g(N)$ (see note on page 47). Thus, the search in an unordered array has an asymptotic complexity linear in the size of input data.

Consider the quantum analogue of the search algorithm. We formulate a quantum search problem in the following form.

There is specified an unordered set of non-negative integers

$$\{0, 1, 2, \ldots, N - 1\},$$

exactly one of which is the sought number, let us denote such sought number by q_0. To check if any value is appropriate, there is used the following function:

$$F: \{0, 1, 2, \ldots, N - 1\} \to \mathbb{B}, \tag{10.2}$$

which operates in accordance with the rule:

$$F(q) = \begin{cases} 1, & \text{if } q = q_0, \\ 0, & \text{if } q \neq q_0. \end{cases} \tag{10.3}$$

In other words, at the target value, the function F is equal to one, at all other values from the definition area it is equal to zero. The problem is to define q_0 provided that the calculation of $F(q)$ in an arbitrary point is taken as one calculation step.

Let us turn to the quantum representation of the considered problem. To the numbers from the set $\{0, 1, 2, \ldots, N - 1\}$ we assign to the quantum states $|0\rangle$, $|1\rangle$, $\ldots$, $|N - 1\rangle$. Checking to see if some state is sought is done with the specified quantum circuit $\mathcal{O}$. This circuit is called **oracle**. It operates in accordance with the following rule:

$$\mathcal{O}|q\rangle = \begin{cases} -|q\rangle, & \text{if } q = q_0, \\ |q\rangle, & \text{if } q \neq q_0. \end{cases} \tag{10.4}$$

Note that with the help of the quantum oracle it is possible to change the phase of the sought state without affecting all other states of the set.

However, phase change alone will not allow us to make a conclusion about q_0 at the macroscopic level. For this purpose, it is necessary to change the observed probabilities inherent in a quantum system, which is what is performed in the **Grover's**[1] **algorithm**, or **quantum search algorithm**.

Grover's algorithm consists of the following steps [3, 4]:

1. Applying Hadamard transform to register $\left| \underbrace{00\ldots0}_{n \text{ qubits}} \right\rangle$ of basis states

$$|00\ldots0\rangle \rightarrow \frac{1}{\sqrt{N}} \sum_{q=0}^{N-1} |q\rangle. \tag{10.5}$$

We denote the obtained state through $|\Psi_0\rangle$.

2. Application of a unitary operator $\mathcal{U} = HSH\mathcal{O}$ to the current state, where H is the Hadamard operator, $S = 2|0\rangle\langle0| - I$ is the so-called **projection operator**. This operator acts in accordance with the rule:

$$S|q\rangle = \begin{cases} |q\rangle, & \text{if } q = 0, \\ -|q\rangle, & \text{if } q \neq 0, \end{cases} \tag{10.6}$$

reversing the phase sign for all states except zero.

3. Step 2 is repeated $m = \left\lfloor \frac{\pi}{4}\sqrt{N} \right\rfloor$ times. Here is used the following designation: $\lfloor x \rfloor$ is an integer part of the real number x, i.e., the largest integer, smaller, or equal to x:

$$\lfloor x \rfloor : \mathbb{R} \rightarrow \mathbb{Z}, \quad \lfloor x \rfloor = \max(n \in \mathbb{Z}, \ n \leqslant x). \tag{10.7}$$

---

[1]Lov Kumar Grover (born 1961).

4. The register is measured in a computational basis.

Note that the properties of the matrix of the operator S are given in Exercise 17. Let us prove the correctness of Grover's algorithm.

In step 1, the register is prepared in the state

$$|\Psi_0\rangle = \frac{1}{\sqrt{N}} (|0\rangle + |1\rangle + \cdots + |q_0\rangle + \cdots + |N - 1\rangle)$$

$$= \frac{1}{\sqrt{N}} |q_0\rangle + \frac{1}{\sqrt{N}} \sum_{q \neq q_0} |q\rangle . \tag{10.8}$$

We denote the coefficients in the obtained linear combination using the auxiliary real-valued parameter θ_0 [5]:

$$\frac{1}{\sqrt{N}} = \sin \theta_0, \quad \frac{\sqrt{N - 1}}{\sqrt{N}} = \cos \theta_0. \tag{10.9}$$

Vector $|\Psi_0\rangle$ we present as

$$|\Psi_0\rangle = \sin \theta_0 |q_0\rangle + \cos \theta_0 |p\rangle , \tag{10.10}$$

where $|p\rangle = \frac{1}{\sqrt{N - 1}} \sum_{q \neq q_0} |q\rangle$ is the normalized superposition of all system states, not including the sought state.

The second step is the conversion of $|\Psi_0\rangle \rightarrow (HSH\mathcal{O}) |\Psi_0\rangle$. Consider separately the HSH operator. According to the definition of S, the following equality is performed for this product

$$HSH = 2H |0\rangle \langle 0| H - HIH = 2 |\Psi_0\rangle \langle \Psi_0| - I. \tag{10.11}$$

Let us prove the supporting statement.

Lemma. *The result of a m-time application of the operator $\mathcal{U}$ to the state $|\Psi_0\rangle$ can be written as follows:*

$$\mathcal{U}^m |\Psi_0\rangle = \sin \theta_m |q_0\rangle + \cos \theta_m |p\rangle , \tag{10.12}$$

where the real constant θ_m equals

$$\theta_m = \theta_0 + m \arcsin \frac{2\sqrt{N - 1}}{N}. \tag{10.13}$$

Proof of the Lemma.

We introduce the auxiliary vector $|\Psi_k\rangle = \sin \theta_k |q_0\rangle + \cos \theta_k |p\rangle$, where $k = 1, 2, \ldots, m$. The ket vector $|\Psi_k\rangle$ in its physical sense represents the result of applying the operator $\mathcal{U}$ to $|\Psi_0\rangle$ exactly k times.

Using the method of mathematical induction, we will prove that the formula (10.12) is valid for all natural m. To do this, we calculate the action of the single operator $\mathcal{U}$ on $|\Psi_k\rangle$.

$$\mathcal{U} |\Psi_k\rangle = \sin \theta_k \mathcal{U} |q_0\rangle + \cos \theta_k \mathcal{U} |p\rangle$$

$$= \sin \theta_k HSH \mathcal{O} |q_0\rangle + \cos \theta_k HSH \mathcal{O} |p\rangle$$

$$= - \sin \theta_k HSH |q_0\rangle + \cos \theta_k HSH |p\rangle$$

$$= - \sin \theta_k (2 |\Psi_0\rangle \langle \Psi_0| - I) |q_0\rangle + \cos \theta_k (2 |\Psi_0\rangle \langle \Psi_0| - I) |p\rangle$$

$$= - \sin \theta_k (2 \langle \Psi_0|q_0\rangle |\Psi_0\rangle - |q_0\rangle) + \cos \theta_k (2 \langle \Psi_0|p\rangle |\Psi_0\rangle - |p\rangle)). \tag{10.14}$$

Then we use the equality (10.10).

$$\mathcal{U} |\Psi_k\rangle = - \sin\theta_k [2\sin\theta_0 (\sin\theta_0 |q_0\rangle + \cos\theta_0 |p\rangle) - |q_0\rangle]$$
$$+ \cos\theta_k [2\cos\theta_0 (\sin\theta_0 |q_0\rangle + \cos\theta_0 |p\rangle) - |p\rangle]. \qquad (10.15)$$

After performing simple trigonometric transformations, we obtain

$$\mathcal{U} |\Psi_k\rangle = |q_0\rangle (\sin\theta_k \cos(2\theta_0) + \cos\theta_k \sin(2\theta_0))$$
$$+ |p\rangle (\cos\theta_k \cos(2\theta_0) - \sin\theta_k \sin(2\theta_0))$$
$$= \sin(\theta_k + 2\theta_0) |q_0\rangle + \cos(\theta_k + 2\theta_0) |p\rangle . \qquad (10.16)$$

Thus, each subsequent application of the $\mathcal{U}$ operator results in adding the constant $2\theta_0$ to the trigonometric function argument in (10.16). The lemma is proved.

From the formula (10.12), you can see that the amplitude of probability of detection of the sought state $|q_0\rangle$ as a result of the measurement procedure is changed by the harmonic law. Depending on the parameter m, the amplitude and therefore the probability may increase or decrease.

The probability of obtaining the sought value is maximized when the function $\sin\theta_m$ takes the maximal value, i.e., is closest to one. This is done under the condition $\theta_m = \pi/2$. Of course, the number of iterations of the algorithm must be integer, therefore $m = \left\lfloor \frac{\pi}{4}\sqrt{N} \right\rfloor$, which is used in step 4 of the quantum search algorithm. In this case, using $\theta_0 = O\left(\frac{1}{\sqrt{N}}\right)$, we get:

$$\theta_m = \theta_0 + \left\lfloor \frac{\pi}{4}\sqrt{N} \right\rfloor \arcsin \frac{2\sqrt{N-1}}{N} = \frac{\pi}{2} + O\left(\frac{1}{\sqrt{N}}\right). \qquad (10.17)$$

This proved the correctness of the Grover's algorithm.

Let us show that the probability of an erroneous result of Grover's algorithm satisfies the asymptotic estimate $O(1/N)$.

Solution.

The probability p_{err} of obtaining a result other than $|q_0\rangle$ when measuring register is equal to the square of the amplitude of the state $|p\rangle$:

$$p_{\text{err}} = \cos^2\theta_m, \qquad (10.18)$$

where θ_m is defined by the formula (10.13). From the definition θ_m follows an asymptotic equality $\theta_m = \theta_0 + O(1/\sqrt{N})$. We obtain

$$p_{\text{err}} = \cos^2\left[\frac{\pi}{2} + O\left(\frac{1}{\sqrt{N}}\right)\right] = O\left(\frac{1}{N}\right). \qquad (10.19)$$

As can be seen from the resulting estimate p_{err}, the probability of error is reduced at $N \to \infty$ by hyperbolic law $\sim 1/N$. $\square$

Thus, the asymptotic complexity of the quantum search algorithm is equal to $O(\sqrt{N})$ and much less than its classical analogue. Nevertheless, it should be noted that to solve the problem of searching the target element in *ordered* data structures, there are known effective classical algorithms (binary search algorithm with a

$O(\log_2 N)$ complexity on average, interpolation search with a $O(\log_2 \log_2 N)$ complexity on average). Compiling quantum analogues of these algorithms is a complex, not fully solved at the present time problem.

References

1. Knuth DE (1998) The art of computer programming. Sorting and searching, vol 3, 2nd edn. Addison-Wesley, Reading, Massachusetts
2. McConnell JJ (2001) Analysis of algorithms: an active learning approach. Jones and Bartlett Publishers, Sudbury, Massachusetts
3. Grover LK (1996) A fast quantum mechanical algorithm for database search. In: Proceedings of the twenty-eighth annual ACM symposium on theory of computing (STOC'96), pp 212–219
4. Grover LK (1996) Quantum mechanics helps in searching for a needle in a haystack. Phys Rev Lett 79(2):325–328
5. Boyer M, Brassard G, Høyer P, Tapp A (1998) Tight bounds on quantum searching. Fortschritte Physik 4—5:493—505

Review Questions

1. Define bit and qubit.
2. What states are referred to the states of computational basis?
3. How is the scalar product operation of two complex vectors introduced?
4. Write down the orthonormality condition of the vector system.
5. What is a gate?
6. Please, explain how the quantum state measurement is performed.
7. List the main elements of a quantum computer.
8. What role do registers play in a quantum computer?
9. Write down Pauli and Dirac matrices.
10. Define the commutator and the anticommutator of the two matrices.
11. Write the Jacobi identity for commutators.
12. Which matrix is called a block matrix?
13. How is the product of operators $U_1, U_2, ..., U_n$ defined?
14. Give definition of quantum circuit.
15. Specify the main operations with one qubit.
16. Write down the matrix representation of Hadamard operator.
17. How the quantum circuits represent the result of the state measurement?
18. What quantum operations are applied to a two qubit system?
19. Formulate the theorem of universal set of elements.
20. Specify the states forming the Bell basis.
21. Give the definition of an entangled quantum state.
22. Explain the nonlocality property of quantum theory.
23. Why is Bell state also referred to as an EPR pair?
24. Formulate the no-cloning theorem.
25. What is the quantum teleportation procedure?
26. Draw a quantum circuit, with which a quantum teleportation can be performed.
27. Specify the main Boolean algebra operations.
28. List the methods of Boolean function defining.
29. Is it possible to build a disjunctive normal form for any Boolean function?
30. Which set of Boolean functions is called universal for classical computations?
31. List the elements of Zhegalkin basis.
32. How the Sheffer stroke works?
33. What is the difference between a Toffoli gate and a Fredkin gate?
34. Describe how the gates of controlled quantum operations work.
35. Draw quantum circuits that implement Toffoli and Fredkin gates using only two-qubit gates.
36. List the techniques for constructing the algebraic normal form of the Boolean function.
37. Formulate an algorithm for constructing a quantum circuit to implement a Boolean function in an algebraic normal form.
38. How is the classical discrete Fourier transform defined?
39. Describe the Fast Fourier Transform algorithm.
40. What is quantum discrete Fourier transform?

41. How many elementary gates need to be used to construct a Quantum Fourier Transform circuit?

42. Describe the phase evaluation algorithm.

Exercises

1. Calculate Pauli matrices commutators $[\sigma_1,\sigma_2], [\sigma_2,\sigma_3]$ and $[\sigma_3,\sigma_1]$.
2. Calculate the product of Pauli matrices $\sigma_1\sigma_2\sigma_3$.
3. What is the sum of products of the squares of Pauli matrices $\sigma_1^2 + \sigma_2^2 + \sigma_3^2$?
* 4. Prove a generalization of the **Euler formula** for Pauli matrices σ_1, σ_2, and σ_3:

$$\exp(i\sigma_k\varphi) = I \cos\varphi + i\sigma_k \sin\varphi \quad \text{for } k = 1, 2, 3, \ \varphi \in \mathbb{R}, \qquad (10.20)$$

where $I = \begin{bmatrix} 1 & 0 \\ 0 & 1 \end{bmatrix}$ is the identity matrix of the second order.

5. Prove that for any A, B, and C matrices of the same size the **Leibniz**[2] **identity**

$$[AB, C] \equiv A[B, C] + [A, C]B \qquad (10.21)$$

is valid.

6. Is it true that if $[A, B] = O$ and $[A, C] = O$, then matrices B and C are commutative?
7. The set $\{I, \sigma_1, \sigma_2, \sigma_3\}$ forms the basis in the vector space of complex matrices of size 2×2. Prove that an arbitrary M matrix can be represented as a linear combination of

$$M = a_0 I + a_1\sigma_1 + a_2\sigma_2 + a_3\sigma_3, \qquad (10.22)$$

where a_{0-3} are some complex numbers.

8. Find the coefficients a_0, a_1, a_2, and a_3 in the expansion (10.22) from the previous exercise.
* 9. Calculate the sum of the squares of the coefficient modules a_0, a_1, a_2, and a_3 in the expansion on the Pauli basis (10.22) of some unitary matrix U.
10. What is the square of the Dirac matrix β?
11. Calculate the product of Dirac matrices $\alpha_1\alpha_2\alpha_3\beta$.
12. Calculate tensor products $\sigma_1 \otimes \sigma_3, \ \sigma_3 \otimes \sigma_1, \ \sigma_2 \otimes \sigma_3, \ \sigma_3 \otimes \sigma_2$.
13. Write down Pauli matrices in terms of tensor products using computational basis vectors $|0\rangle$ and $|1\rangle$.
14. Which of the following matrices are Hermitian?

(1) $\begin{bmatrix} 2 & 2+i \\ 2-i & -2 \end{bmatrix}$;

(2) $\begin{bmatrix} 1 & 0 \\ 0 & 0 \end{bmatrix}$;

(3) $\dfrac{1}{\sqrt{5}} \begin{bmatrix} 2i & 3i \\ -3i & 2i \end{bmatrix}$;

(4) $\begin{bmatrix} 0 & 1 & -1 \\ 0 & -1 & 1 \end{bmatrix}$;

---

[2]Gottfried Wilhelm Leibniz (1646–1716).

(5) $\begin{bmatrix} 0 & -i \\ -i & 0 \end{bmatrix}$;

(6) $\dfrac{1}{\sqrt{3}} \begin{bmatrix} 1 & 0 & 2 \\ 0 & -i & 0 \\ 2 & 0 & i \end{bmatrix}$;

(7) $\begin{bmatrix} 1 & 1 & 1 & 1 \\ 1 & 1 & -i & i \\ 1 & i & 1 & -1 \\ 1 & -i & -1 & 1 \end{bmatrix}$;

(8) $\begin{bmatrix} 1 & 1 & 1 & 1 \\ 1 & 1 & -1 & 1 \\ 1 & -1 & 1 & -1 \\ 1 & -1 & -1 & 1 \end{bmatrix}$.

15. Which of the following matrices are unitary?

(1) $\begin{bmatrix} 1 & 2 \\ 2 & 3 \end{bmatrix}$;

(2) $\dfrac{1}{\sqrt{5}} \begin{bmatrix} 2 & i \\ i & 2 \end{bmatrix}$;

(3) $\begin{bmatrix} 1 & 2 & 3 \\ 3 & 2 & 1 \end{bmatrix}$;

(4) $\dfrac{1}{3} \begin{bmatrix} 2+i & 2i \\ -2i & -2+i \end{bmatrix}$;

(5) $\dfrac{1}{\sqrt{10}} \begin{bmatrix} 3 & i \\ -i & 3 \end{bmatrix}$;

(6) $\dfrac{1}{\sqrt{\cosh(2\pi)}} \begin{bmatrix} \cosh \pi & -i \sinh \pi \\ -i \sinh \pi & \cosh \pi \end{bmatrix}$;

(7) $\dfrac{1}{\sqrt{3}} \begin{bmatrix} e^{-3i} & 0 & 0 \\ 0 & 1 & 0 \\ 0 & 0 & e^{3i} \end{bmatrix}$;

(8) $\dfrac{1}{2} \begin{bmatrix} 1 & 1 & 1 & 1 \\ 1 & -i & -1 & i \\ 1 & -1 & 1 & -1 \\ 1 & i & -1 & -i \end{bmatrix}$.

16. The elements of matrix $M(\theta)$ depend on the real-valued parameter θ:

$$M = \begin{bmatrix} \frac{1}{2}(1+\cos\theta) & \frac{1}{\sqrt{2}}\sin\theta & \frac{1}{2}(-1+\cos\theta) \\ -\frac{1}{\sqrt{2}}\sin\theta & \cos\theta & -\frac{1}{\sqrt{2}}\sin\theta \\ \frac{1}{2}(-1+\cos\theta) & \frac{1}{\sqrt{2}}\sin\theta & \frac{1}{2}(1+\cos\theta) \end{bmatrix}. \tag{10.23}$$

Determine if the matrix $M(\theta)$ is Hermitian? unitary?

17. Prove that the **quantum inversion** matrix of size $N \times N$, which occurs while analyzing the search algorithm in a quantum database, is unitary:

$$D_N = \begin{bmatrix} -1+\frac{2}{N} & \frac{2}{N} & \cdots & \frac{2}{N} \\ \frac{2}{N} & -1+\frac{2}{N} & \cdots & \frac{2}{N} \\ & & \cdots & \\ \frac{2}{N} & \frac{2}{N} & \cdots -1+\frac{2}{N} \end{bmatrix} \tag{10.24}$$

for all $N > 1$.

18. Find all complex Hermitian matrices commuting with the real matrix $\begin{bmatrix} 1 & 1 \\ 1 & 1 \end{bmatrix}$.

* 19. Find all complex unitary matrices commuting with the real matrix $\begin{bmatrix} 1 & 1 \\ 1 & 1 \end{bmatrix}$.

* 20. Let U be arbitrary unitary matrix of size 2×2. Is there a matrix V with the property $V^2 = U$?

21. How many real parameters are necessary and sufficient to set the state of one qubit?

22. The **fidelity** of two states described by vectors $|\psi_1\rangle$ and $|\psi_2\rangle$, is called a real number $\mathcal{F} = |\langle\psi_1|\psi_2\rangle|^2$.

(1) Prove that $F = 1$ if the states $|\psi_1\rangle$ and $|\psi_2\rangle$ are identical, and $F = 0$ if the states are orthogonal.

(2) Calculate the fidelity of single-qubit states

$$|\psi_1\rangle = \frac{1}{\sqrt{2}}[|0\rangle + e^{i\varphi_1}|1\rangle], \tag{10.25}$$

$$|\psi_2\rangle = \frac{1}{\sqrt{2}}[|0\rangle + e^{i\varphi_2}|1\rangle], \tag{10.26}$$

where $0 \leqslant \varphi_{1,2} \leqslant 2\pi$.

23. Calculate the fidelity of single-qubit states

$$|\psi_1\rangle = e^{i\gamma_1}[\cos(\theta_1/2)|0\rangle + e^{i\varphi_1}\sin(\theta_1/2)|1\rangle], \tag{10.27}$$

$$|\psi_2\rangle = e^{i\gamma_2}[\cos(\theta_2/2)|0\rangle + e^{i\varphi_2}\sin(\theta_2/2)|1\rangle], \tag{10.28}$$

where $0 \leqslant \theta_{1,2} \leqslant \pi, 0 \leqslant \varphi_{1,2}, \gamma_{1,2} < 2\pi$.

24. Prove **Schwarz**[3] **inequality** (which is also called **Cauchy**[4]**–Bunyakovsky**[5]**–Schwarz inequality**):
    for arbitrary vectors $|x\rangle$, $|y\rangle \in \mathbb{H}$ the inequality

    $$|\langle x|y\rangle|^2 \leqslant \langle x|x\rangle \langle y|y\rangle \qquad (10.29)$$

    is valid. The equality in (10.29) holds if and only if the vectors $|x\rangle$ and $|y\rangle$ differ by a scalar multiplier, i.e., the vectors are proportional.

25. Prove that if the states $|\psi_1\rangle$ and $|\psi_2\rangle$ are orthogonal, then the **Pythagoras**[6] **theorem**

    $$\langle \psi_1 + \psi_2|\psi_1 + \psi_2\rangle = \langle \psi_1|\psi_1\rangle + \langle \psi_2|\psi_2\rangle \qquad (10.30)$$

    is valid.

26. Write the state $|\psi\rangle$ of the quantum system as a vector-column corresponding to the computational basis:

    (1) $|\psi\rangle = |010\rangle$;
    (2) $|\psi\rangle = |101\rangle$.

27. The considered quantum system is formed by n interacting qubits. Write the state $|\psi\rangle$ of the quantum system in the form of a vector-column corresponding to the computational basis:

    (1) $|\psi\rangle = |00\ldots011\rangle$;
    (2) $|\psi\rangle = |11\ldots110\rangle$.

28. A student conducting an experiment in a chemical laboratory accidentally filled a page of a quantum computing summary with reagent, on which a vector of some three-qubit state $|\psi\rangle$ was written. As a result, we were unable to read some symbols. If you mark such symbols "$*$", the record will look as follows:

    $$|\psi\rangle = \frac{1}{10}(|{*}00\rangle + 2\,|001\rangle - 4\,|010\rangle + i\,|011\rangle$$
    $$+\, i\,|100\rangle - |101\rangle + |110\rangle + 5\sqrt{*}\,|{*}11\rangle). \qquad (10.31)$$

    Recover the state vector $|\psi\rangle$ written in the compendium if one character "$*$" matches exactly one original character.

29. Let $|\psi\rangle = (|0\rangle - i\,|1\rangle)/\sqrt{2}$. Compute the tensor product $|\psi\rangle\,|\psi\rangle$.

30. Determine whether the basis

    $$\left\{ \frac{1}{10}\begin{bmatrix} 1 \\ 3i \end{bmatrix},\; \frac{1}{10}\begin{bmatrix} 3i \\ 1 \end{bmatrix} \right\} \qquad (10.32)$$

    is orthonormalized.

---

[3] Karl Hermann Amandus Schwarz (1843–1921).
[4] Augustin-Louis Cauchy (1789–1857).
[5] Victor Yakovlevich Bunyakovsky (1804–1889).
[6] Pythagoras of Samos (Pythagoras of Samos, Πνθαγόρας ὁ Σάμιος) (ca. 570 B.C.—ca. 495 B.C.).

31. Determine what is the result of the measurement of the state

$$|\psi\rangle = \frac{1+i}{6}|0\rangle + \frac{3+5i}{6}|1\rangle \tag{10.33}$$

in the computational basis $\{|0\rangle, |1\rangle\}$?

32. Determine what is the result of the measurement of the state

$$|\psi\rangle = \frac{5+i}{6}|0\rangle + \frac{1-3i}{6}|1\rangle \tag{10.34}$$

in the basis $B_1 = \{|0\rangle, |1\rangle\}$ and $B_2 = \{|+\rangle, |-\rangle\}$, where $|+\rangle = \frac{1}{2}(|0\rangle + |1\rangle)$, $|-\rangle = \frac{1}{2}(|0\rangle - |1\rangle)$?

33. What is the result of the measurement of the state

$$|\psi\rangle = \frac{2+3i}{7}|0\rangle + \frac{6}{7}|1\rangle \tag{10.35}$$

in the basis $B_1 = \{|0\rangle, |1\rangle\}$ and $B_2 = \{|+\rangle, |-\rangle\}$, where $|+\rangle = \frac{1}{2}(|0\rangle + |1\rangle)$, $|-\rangle = \frac{1}{2}(|0\rangle - |1\rangle)$?

34. What is the result of the measurement of the state

$$|\psi\rangle = (1-i)\frac{7}{10}|0\rangle + (1+i)|1\rangle \tag{10.36}$$

in the basis $B_1 = \{|0\rangle, |1\rangle\}$ and $B_2 = \{|+\rangle, |-\rangle\}$, where $|+\rangle = \frac{1}{2}(|0\rangle + |1\rangle)$, $|-\rangle = \frac{1}{2}(|0\rangle - |1\rangle)$?

35. Calculate the result of the quantum circuit action

$$-\boxed{X}\qquad-\boxed{H}\qquad-\boxed{S}\qquad-\boxed{H}-$$

on the qubit, which is initially in the state $|0\rangle$.

36. Calculate the result of the quantum circuit action

$$-\boxed{Y}\qquad-\boxed{H}\qquad-\boxed{Z}\qquad-\boxed{T}-$$

on the qubit, which is initially in the state $|1\rangle$.

37. Calculate the result of the quantum circuit action

$$-\boxed{T}\qquad-\boxed{H}\qquad-\boxed{S}\qquad-\boxed{H}\qquad-\boxed{S}-$$

on the qubit, which is initially in the state $|\psi\rangle = u|0\rangle + v|1\rangle$, where $u, v \in \mathbb{C}$.

38. Determine what will be equal to the measurement result in a computational basis $\{|0\rangle, |1\rangle\}$:

$$|0\rangle -\boxed{S}\qquad-\boxed{H}\qquad-\boxed{Z}\qquad-\boxed{H}\qquad-\boxed{\measuredangle}$$

39. Consider a two-qubit system brought to the state $|\Psi\rangle = \frac{1}{\sqrt{2}}\big(|0\rangle|\psi_1\rangle + |1\rangle|\psi_2\rangle\big)$, where $|\psi_1\rangle$ and $|\psi_2\rangle$ are some normalized one-qubit quantum states. Apply Hadamard operation to the first qubit and perform a measurement of the first qubit. What is the result of this measurement?

40. It is known that a two-qubit system is set to the state $|\psi\rangle = \sqrt{\frac{3}{5}}|10\rangle + i\sqrt{\frac{2}{5}}|11\rangle$. A CNOT operation is performed over the system, after which Hadamard operation is applied to the second (controlled) qubit. What state will the system be in?

41. What is the result of Hadamard gate action on the basis state $|q\rangle$, where $q \in \{0, 1\}$?

42. Let us input the CNOT element state $|\psi_{in}\rangle = (u|0\rangle + v|1\rangle)|1\rangle$. What is the output state?

43. Show that the quantum circuit

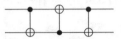

converts the state $|\psi_1\psi_2\rangle$ to $|\psi_2\psi_1\rangle$.

44. Implement a CNOT gate using only Hadamard gate and controlled Z.

45. Implement the Fredkin gate using a quantum circuit that uses three Toffoli gates.

46. Check if the set $\{\vee, \wedge, 1\}$ is universal.

* 47. Is it possible to implement a Toffoli gate using a quantum circuit that uses only Fredkin gates without auxiliary qubits?

48. Consider the transformation to which corresponds the following quantum circuit:

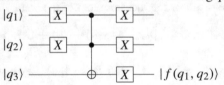

What Boolean function of two arguments $f(q_1, q_2)$ with condition $q_3 = 0$ does this circuit implement?

49. How many elementary gates contains the quantum circuit from Example 6.2 on page 33 that implements operation $\underbrace{CC\ldots C}_{n \text{ times}} NOT$?

50. Write the result of applying

   (1) Toffoli gate;
   (2) Fredkin gate

to the state $|\psi\rangle = \frac{1}{2}(|000\rangle + |001\rangle + |101\rangle - |111\rangle)$.

51. Suggest a circuit that implements an arbitrary unitary operation U controlled by two qubits:

52. At the exam on the course of quantum computing the student is offered to analyze the following circuit, the input to which is submitted the state $|01\rangle$. Remembering that the Hadamard transformation H coincides with the inverse transformation

$H \times H = I$ (see Example 3.2 on page 12), the student concludes that the state of the first qubit does not change.

Is the student right?

53. It is known that the two-qubit system is brought to the Bell state (4.1). Calculate the result of the conversion X applied to each of the two qubits.

54. It is known that the two-qubit system is brought to the Bell state (4.1). Calculate the result of the conversion Z, applied to each of the two qubits.

55. Define the criterion (i.e., the necessary and sufficient condition) that the state of the two-qubit system $\alpha |00\rangle + \beta |01\rangle + \gamma |10\rangle + \delta |11\rangle$ is presented as a product of $|\psi_1\rangle |\psi_2\rangle$ some states $|\psi_1\rangle$ and $|\psi_2\rangle$.

56. Show that the Bell states $|\beta_1\rangle - |\beta_4\rangle$ form an orthonormalized basis.

57. On an exam for a quantum computing course, a student states that a single qubit $\psi = u |0\rangle + v |1\rangle$ can easily be copied using the controlled NOT and auxiliary $|0\rangle$:

Is the student right?

58. Prove that **GHZ-state**[7] $|GHZ\rangle = \dfrac{1}{\sqrt{2}}(|000\rangle + |111\rangle)$ is entangled [1]

59. Suggest a quantum circuit which, based on the fiduciary state $|000\rangle$, forms the state $|GHZ\rangle = \dfrac{1}{\sqrt{2}}(|000\rangle + |111\rangle)$.

60. Suggest a quantum circuit which, based on the state $\left| \underbrace{00\ldots0}_{n \text{ qubits}} \right\rangle$ forms the state

$$\frac{1}{\sqrt{2}}(|00\ldots0\rangle + |11\ldots1\rangle).$$

61. Prove that the four-qubit state

$$|\psi\rangle = \frac{1}{2}(|0001\rangle + |0010\rangle + |1100\rangle + |1111\rangle)$$

is entangled.

62. The set of qubits is brought to the state

$$|\psi\rangle = \frac{1}{5}(2|000\rangle + 2\sqrt{2}|010\rangle + 2|101\rangle + 3|111\rangle). \qquad (10.37)$$

With what probability the result of measuring the second qubit will be equal to $|0\rangle$?

63. Determine which of the following states are entangled:

---

[7]Quantum state is named after the first letters of the names of physicists Daniel M. Greenberger (born 1932), Michael A. Horne and Anton Zeilinger (born 1945).

(1) $|\psi_1\rangle = \dfrac{1}{\sqrt{2}}(|00\rangle + i\,|01\rangle)$;

(2) $|\psi_2\rangle = \dfrac{1}{2}(|00\rangle + |01\rangle + |10\rangle + |11\rangle)$;

(3) $|\psi_3\rangle = \dfrac{1}{3\sqrt{2}}|00\rangle - \dfrac{5}{\sqrt{6}}|01\rangle + \dfrac{1}{12}|10\rangle - \dfrac{1}{4}\sqrt{\dfrac{5}{3}}|11\rangle$;

(4) $|\psi_4\rangle = \dfrac{3}{7}|00\rangle - \dfrac{3\sqrt{33}}{28}|01\rangle - \dfrac{1}{\sqrt{7}}|10\rangle + \dfrac{33}{4\sqrt{7}}|11\rangle$.

64. Two states are given:

$$|\psi_1\rangle = \dfrac{1}{2}(-\,|00\rangle + |01\rangle + |10\rangle + |11\rangle));$$

$$|\psi_2\rangle = \dfrac{1}{2}(-\,|00\rangle + |01\rangle - |10\rangle + |11\rangle)).$$

Determine which one is entangled and which one is not entangled.

65. Build ANF for the function $f(x_1, x_2)$, defined by the vector of values $= (1011)$.

66. Using the undefined coefficients method, build Zhegalkin polynomial using the vector of function values:

(1) $\alpha_g = (0011\ 0110)$;
(2) $\alpha_h = (1100\ 1100\ 0011\ 0000)$.

67. Using the equivalent transformation method, build Zhegalkin polynomial for each of the following functions:

(1) $g(x_1, x_2, x_3) = (x_1 \wedge x_2) \oplus (x_1 \downarrow x_3)$;
(2) $h(x_1, x_2, x_3, x_4) = \overline{(x_1 \rightarrow x_2)} \rightarrow (x_3 \leftrightarrow x_4)$.

68. Using any of the known methods, build an algebraic normal form for each of the following Boolean functions:

(1) $f(x_1, x_2, x_3) = (x_1 x_2 x_3) \vee (\overline{x}_1 \overline{x}_2 \overline{x}_3)$;
(2) $g(x_1, x_2, x_3) = (x_1 \vee x_2 \vee x_3)(\overline{x}_1 \vee \overline{x}_2 \vee \overline{x}_3)$.

*   69. Using any of the known methods, build an algebraic normal form for each of the following Boolean functions:

(1) $f(x_1, x_2) = x_1 \rightarrow x_2$;
(2) $f(x_1, x_2, x_3) = (x_1 \rightarrow x_2) \rightarrow x_3$;
(3) $f(x_1, x_2, x_3, x_4) = ((x_1 \rightarrow x_2) \rightarrow x_3) \rightarrow x_4$;
(4) $f(x_1, x_2, x_3, x_4, x_5) = (((x_1 \rightarrow x_2) \rightarrow x_3) \rightarrow x_4) \rightarrow x_5$.

70. Determine which function $P \colon [0, 7] \rightarrow \mathbb{Z}_0$ is implemented by the following quantum circuit:

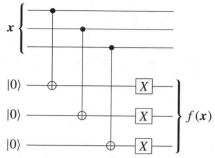

71. Make a table of values of the function $g(x)$ implemented using the following quantum circuit:

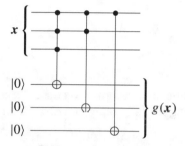

72. Construct a quantum circuit that implements the function $f(x) = (x + 2)^2$. Assume that the domain of the function is the set $\{0, 1, 2, \ldots, 7\}$.

73. Create a quantum circuit that matches the number of the planet of the Solar System number of its satellites.

74. Let us denote the roots of the equation $z^N = 1$, where N is a natural number, by $\omega_k, k = 0, \ldots, N - 1$ (see page 46). Prove that the following statements are correct [2]:

(a) on the complex plane, the points corresponding to ω_k values are located at the tops of the correct N-angle inscribed in the unit circle which center is located at the beginning of coordinates;

(b) $\omega_{k+n/2} = -\omega_k$ for even N and $0 \leqslant k \leqslant n/2 - 1$;

(c) $\sum_{k=0}^{N-1} \omega_k = 0$ for $N > 1$;

(d) $\prod_{k=0}^{N-1} \omega_k = (-1)^{N-1}$.

75. Prove the validity of identities for the roots of the unity ω_k, where $0 \leqslant k \leqslant N-1$, for all natural values N:

(a) $\prod_{k=0}^{N-1} (z - \omega_k) = z^N - 1$;

(b) $\displaystyle\sum_{k=0}^{N-1} (\omega_k)^d = \begin{cases} 0, & 1 \leqslant d \leqslant N-1; \\ N, & d = N. \end{cases}$

* 76. Prove the validity of identities for the roots of unity of Nth degree ω_k, where $k = 0, 1, \ldots, N-1$, for all values $N > 2$:

(1) $\displaystyle\sum_{k=0}^{N-2} \omega_k \omega_{k+1} = -\omega_{N-1};$

(2) $\displaystyle\sum_{k=1}^{N-2} \frac{\omega_{k-1}\omega_{k+1}}{\omega_k} = -(1 + \omega_{N-1});$

(3) $\displaystyle\sum_{\substack{k,k'=0 \\ k<k'}}^{N-1} \omega_k \omega_{k'} = 0;$

(4) $\displaystyle\sum_{\substack{k,k'=0 \\ k<k'}}^{N-1} \frac{\omega_k \omega_{k'}}{\omega_{k'-k}} = \frac{N}{1 - \omega_2}.$

77. Let $\mathbf{y}$ be a discrete Fourier transform of vector $\mathbf{x}$ of length N. Prove the **Parseval**[8] **theorem**, connecting the full energy of a discrete signal in time and frequency domains:

$$\sum_{k=0}^{N-1} |x_k|^2 = \frac{1}{N} \sum_{k=0}^{N-1} |y_k|^2. \tag{10.38}$$

78. Using the discrete Fourier transform definition, calculate $\mathcal{F}[\mathbf{x}]$, if:

(1) $\mathbf{x} = (0, 1, 0, 1);$
(2) $\mathbf{x} = (1, 0, 1, 0).$

79. Calculate the DFT of the vector $\mathbf{x}$ of length N with components equal to the binomial coefficients $x_n = C_{N-1}^n$, where $0 \leqslant n \leqslant N-1$ and the binomial coefficients $C_p^q = \dfrac{p!}{q!(p-q)!}$ are denoted by C_p^q.

80. Calculate the DFT of the vector $\mathbf{x}$ of length $N = 2^m$ for some integer positive m, if

(1) $\mathbf{x} = (\underbrace{0, 0, \ldots, 0, 0,}_{N/2 \text{ zeros}} \underbrace{1, 1, \ldots, 1, 1}_{N/2 \text{ units}});$

(2) $\mathbf{x} = (\underbrace{1, 1, \ldots, 1, 1,}_{N/2 \text{ units}} \underbrace{0, 0, \ldots, 0, 0}_{N/2 \text{ zeros}}).$

81. Prove that Hadamard gate implements one-qubit Fourier transform.

---

[8]Parseval (Marc-Antoine Parseval des Chênes) (1755–1836).

82. Calculate the Quantum Fourier transform of n-qubit state $|00\ldots0\rangle$.

83. At the input of the Fourier quantum discrete transform circuit let us give a superposition of states with equal coefficients $|\psi\rangle = \dfrac{1}{\sqrt{2^n}} \sum_{q=0}^{2^n-1} |q\rangle$. What is the state of the output of this circuit?

84. Calculate the result of the action of the Fourier quantum transform on n-qubit state

$$|\psi\rangle = \frac{1}{\sqrt{2^{n-1}}} \sum_{q=0}^{N-1} \cos \frac{2\pi q}{N} |q\rangle . \tag{10.39}$$

85. Suggest a quantum circuit to implement the inverse Fourier transform $\mathcal{F}^{-1}[y] = x$:

$$x_n = \frac{1}{N} \sum_{k=0}^{N-1} \omega^{-kn} y_k, \quad 0 \leqslant n \leqslant N - 1.$$

86. Calculate the speedup achieved by the Quantum Fourier Transform algorithm compared to the classical Fast Fourier Transform algorithm.

87. Calculate the speedup achieved by the Quantum Fourier Transform algorithm compared to the parallel version of the Fast Fourier Transform algorithm presented on page 47.

Answers, Hints, and Solutions

1. *Answer.* $[\sigma_1, \sigma_2] = 2i\sigma_3$, $[\sigma_2, \sigma_3] = 2i\sigma_1$, $[\sigma_3, \sigma_1] = 2i\sigma_2$.
2. *Solution.* $\sigma_1\sigma_2\sigma_3 = iI$, where I is the identity matrix of size 2×2.
3. *Answer.* $\sigma_1^2 + \sigma_2^2 + \sigma_3^2 = 3I$, where I is the identity matrix of the second order.
4. *Solution.*

We will use the Taylor[9] formula to decompose the exponents into a power series:

$$e^t = \sum_{j=0}^{\infty} \frac{t^j}{j!}.$$

In this formula, we substitute $t = i\sigma_k\varphi$ for $k \in \{1, 2, 3\}$ and $\varphi \in \mathbb{R}$. We obtain

$$\exp(i\sigma_k\varphi) = \sum_{j=0}^{\infty} \frac{(i\sigma_k\varphi)^j}{j!} = \sum_{j=0}^{\infty} \frac{i^j\sigma_k^j\varphi^j}{j!}.$$

Note that $\sigma_k^2 = I$ (see formula (2.3)), which leads to the validity of $\sigma_k^{2j} = I$ and $\sigma_k^{2j+1} = \sigma_k$ for all non-negative integer j. Next, we present the decomposition in the Taylor series with two sums, by odd and even degrees:

$$\exp(i\sigma_k\varphi) = \sum_{j=0}^{\infty} \left(\frac{i^{2j}\sigma_k^{2j}\varphi^{2j}}{(2j)!} + \frac{i^{2j+1}\sigma_k^{2j+1}\varphi^{2j+1}}{(2j+1)!} \right)$$

$$= \sum_{j=0}^{\infty} \frac{(-1)^j I\varphi^{2j}}{(2j)!} + \sum_{j=0}^{\infty} \frac{(-1)^j i\sigma_k\varphi^{2j+1}}{(2j+1)!}$$

$$= I \sum_{j=0}^{\infty} \frac{(-1)^j\varphi^{2j}}{(2j)!} + i\sigma_k \sum_{j=0}^{\infty} \frac{(-1)^j\varphi^{2j+1}}{(2j+1)!}$$

$$= I \cos\varphi + i\sigma_k \sin\varphi.$$

Obtained sums $\displaystyle\sum_{j=0}^{\infty} \frac{(-1)^j\varphi^{2j}}{(2j)!}$ and $\displaystyle\sum_{j=0}^{\infty} \frac{(-1)^j\varphi^{2j+1}}{(2j+1)!}$ converge to $\cos\varphi$ and $\sin\varphi$, respectively. As a result, we get:

$$\exp(i\sigma_k\varphi) = I \cos\varphi + i\sigma_k \sin\varphi \quad \text{for } k \in \{1, 2, 3\}, \ \varphi \in \mathbb{R}.$$

5. *Proof.*

Convert the right side of the equality (10.21) based on the definition on the page 7:

$$A[B, C] + [A, C]B = A(BC - CB) + (AC - CA)B.$$

Next, we expand the brackets and give the similar summands, and then once again use the definition of the commutator:

$$A[B, C] + [A, C]B = ABC - ACB + ACB - CAB$$

$$= ABC - CAB = [AB, C].$$

---

[9]Brook Taylor (1685–1731).

Thus, the identity (10.21) is proved.

Note that similar transformations easily prove the Leibniz identity in another form:

$$[A, BC] \equiv B[A, C] + [A, B]C.$$

6. *Solution.*

It is not true. The following counterexample can be used as a statement from the problem condition:

$$A = \begin{bmatrix} 1 & 0 \\ 0 & 1 \end{bmatrix}, \quad B = \begin{bmatrix} 1 & 0 \\ 0 & -1 \end{bmatrix}, \quad C = \begin{bmatrix} 0 & 1 \\ 1 & 0 \end{bmatrix}.$$

In this case the equalities $[A, B] = [A, C] = O$ are valid, while

$$[B, C] = \begin{bmatrix} 1 & 0 \\ 0 & -1 \end{bmatrix}\begin{bmatrix} 0 & 1 \\ 1 & 0 \end{bmatrix} - \begin{bmatrix} 0 & 1 \\ 1 & 0 \end{bmatrix}\begin{bmatrix} 1 & 0 \\ 0 & -1 \end{bmatrix} = \begin{bmatrix} 0 & -2 \\ -2 & 0 \end{bmatrix} \neq O.$$

Consequently, B and C matrices are not necessarily commutative.

7. *Proof.*

Denote the elements of matrix M through m_{ij}, $i, j = 1, 2$:

$$M = \begin{bmatrix} m_{11} & m_{12} \\ m_{21} & m_{22}. \end{bmatrix}$$

The expansion (10.22) leads to a system of linear equations relative to the coefficients a_{0-3}:

$$a_0\begin{bmatrix} 1 & 0 \\ 0 & 1 \end{bmatrix} + a_1\begin{bmatrix} 0 & 1 \\ 1 & 0 \end{bmatrix} + a_2\begin{bmatrix} 0 & -i \\ i & 0 \end{bmatrix} + a_3\begin{bmatrix} 1 & 0 \\ 0 & -1 \end{bmatrix} = \begin{bmatrix} m_{11} & m_{12} \\ m_{21} & m_{22} \end{bmatrix},$$

$$\begin{cases} a_0 & + a_3 = m_{11}, \\ a_1 - ia_2 + & = m_{12}, \\ a_1 + ia_2 + & = m_{21}, \\ a_0 & - a_3 = m_{22}. \end{cases}$$

Let us calculate the determinant of the resulting system by applying expansion through low-order determinants:

$$\begin{bmatrix} 1 & 0 & 0 & 1 \\ 0 & 1 & -i & 0 \\ 0 & 1 & i & 0 \\ 1 & 0 & 0 & -1 \end{bmatrix} = 1 \times \begin{bmatrix} 1 & -i & 0 \\ 1 & i & 0 \\ 0 & 0 & -1 \end{bmatrix} - 1 \times \begin{bmatrix} 0 & 1 & -i \\ 0 & 1 & i \\ 1 & 0 & 0 \end{bmatrix}$$

$$= 1 \times (-1) \times \begin{bmatrix} 1 & -i \\ 1 & i \end{bmatrix} - 1 \times 1 \times \begin{bmatrix} 1 & -i \\ 1 & i \end{bmatrix}$$

$$= (-1)(2i) - 1(2i) = -4i \neq 0.$$

The non-zero determinant of the system is a sufficient condition for existence of an unique solution (a_0, a_1, a_2, a_3). Consequently, an arbitrary matrix $M = \begin{bmatrix} m_{11} & m_{12} \\ m_{21} & m_{22} \end{bmatrix}$ can be represented as a linear combination

$$M = a_0 I + \sum_{i=1}^{3} \sigma_i.$$

Note. The set $\{I, \sigma_1, \sigma_2, \sigma_3\}$, where I is the identity matrix of size 2×2, σ_i are Pauli matrices, is called the **Pauli basis** of the vector space of complex matrices of the second order.

8. *Solution.*

Let us represent an arbitrary matrix M in the following form

$$M = \begin{bmatrix} m_{11} & m_{12} \\ m_{21} & m_{22} \end{bmatrix},$$

where $m_{ij} \in \mathbb{C}$ for $i, j \in \{1, 2\}$. With the help of algebraic transformations, we obtain

$$\begin{bmatrix} m_{11} & m_{12} \\ m_{21} & m_{22} \end{bmatrix}$$

$$= \frac{1}{2} \begin{bmatrix} (m_{11} + m_{22}) + (m_{11} - m_{22}) & (m_{12} + m_{21}) + (m_{12} - m_{21}) \\ (m_{12} + m_{21}) - (m_{12} - m_{21}) & (m_{11} + m_{22}) - (m_{11} - m_{22}) \end{bmatrix}$$

$$= \frac{1}{2} \left((m_{11} + m_{22})I + (m_{12} + m_{21})\sigma_1 + \mathrm{i}(m_{12} - m_{21})\sigma_2 + (m_{11} - m_{22})\sigma_3 \right).$$

Thus, the vector of the Pauli basis expansion coefficients is equal to

$$(a_0, a_1, a_2, a_3) = \frac{1}{2} \left((m_{11} + m_{22}), (m_{12} + m_{21}), \mathrm{i}(m_{12} - m_{21}), (m_{11} - m_{22}) \right).$$

9. *Solution.*

First method

Using the unitarity property of matrix U and the ratio for the products of Pauli matrices, write down the following:

$$UU^\dagger = \left(a_0 I + \sum_{i=1}^{3} \sigma_i \right) \left(a_0 I + \sum_{i=1}^{3} \sigma_i \right)^\dagger$$

$$= (a_0 a_0^* + a_1 a_1^* + a_2 a_2^* + a_3 a_3^*)I + (a_0 a_1^* + a_1 a_2^*)\sigma_1 - \mathrm{i}(a_2 a_3^* - a_3 a_2^*)\sigma_1$$

$$+ (a_0 a_3^* + a_3 a_0^*)\sigma_2 + \mathrm{i}(a_1 a_3^* - a_3 a_1^*)\sigma_2 + (a_0 a_3^* + a_3 a_0^*)\sigma_3 - \mathrm{i}(a_1 a_3^* - a_3 a_1^*)\sigma_3.$$

Since matrices $I, \sigma_1, \sigma_2, \sigma_3$ form the basis, the coefficient at I in the expansion $UU^\dagger = I$ is equal to one:

$$a_0 a_0^* + a_1 a_1^* + a_2 a_2^* + a_3 a_3^* = 1.$$

Consequently, the sum of squares of coefficients modules of Pauli basis expansion of unitary matrix is equal to

$$|a_0|^2 + |a_1|^2 + |a_2|^2 + |a_3|^2 = 1.$$

The simplicity of the solution of the exercise in this way manifested itself in the fact that it was not necessary to write down an explicit form of elements of the initial unitary matrix U.

Second method

Components of the vector (a_0, a_1, a_2, a_3) are represented through elements of the source matrix $U = \begin{bmatrix} u_{11} & u_{12} \\ u_{21} & u_{22} \end{bmatrix}$ by formulas:

$$a_0 = \frac{1}{2}(u_{11} + u_{22}), \quad a_1 = \frac{1}{2}(u_{12} + u_{21}),$$

$$a_2 = \frac{i}{2}(u_{12} - u_{21}), \quad a_3 = \frac{1}{2}(u_{11} - u_{22}).$$

Considering this, let us calculate $\sum_{i=0}^{3} |a_i|^2$:

$$\sum_{i=0}^{3} |a_i|^2$$

$$= \frac{1}{2}(u_{11} + u_{22})\left(\frac{1}{2}(u_{11} + u_{22})\right)^* + \frac{1}{2}(u_{12} + u_{21})\left(\frac{1}{2}(u_{12} + u_{21})\right)^*$$

$$= \frac{i}{2}(u_{12} - u_{21})\left(\frac{i}{2}(u_{12} - u_{21})\right)^* + \frac{1}{2}(u_{11} - u_{22})\left(\frac{1}{2}(u_{11} - u_{22})\right)^*$$

$$= \frac{1}{4}\big(|u_{11}|^2 + u_{11}u_{22}^* + u_{11}^*u_{22} + |u_{22}|^2 + |u_{12}|^2 + u_{12}u_{21}^* + u_{12}^*u_{21} + |u_{21}|^2$$

$$+ |u_{12}|^2 - u_{12}u_{21}^* - u_{12}^*u_{21} + |u_{21}|^2 - |u_{11}|^2 - u_{11}u_{22}^* - u_{11}^*u_{22} + |u_{22}|^2\big)$$

$$= \frac{1}{2}(|u_{11}|^2 + |u_{12}|^2 + |u_{21}|^2 + |u_{22}|^2).$$

Due to the unitarity of the matrix U, we have

$$\begin{bmatrix} u_{11} & u_{12} \\ u_{21} & u_{22} \end{bmatrix}\begin{bmatrix} u_{11} & u_{12} \\ u_{21} & u_{22} \end{bmatrix}^{\dagger}\begin{bmatrix} u_{11} & u_{12} \\ u_{21} & u_{22} \end{bmatrix}\begin{bmatrix} u_{11}^* & u_{21}^* \\ u_{12}^* & u_{22}^* \end{bmatrix}$$

$$= \begin{bmatrix} |u_{11}|^2 + |u_{12}|^2 & u_{11}u_{21}^* + u_{12}u_{22}^* \\ u_{11}^*u_{21} + u_{12}^*u_{22} & |u_{21}|^2 + |u_{22}|^2 \end{bmatrix} = I,$$

from where follow the properties of unitary matrix coefficients $|u_{11}|^2 + |u_{12}|^2 = 1$, $|u_{21}|^2 + |u_{22}|^2 = 1$. As a result we obtain

$$\sum_{i=0}^{3} |a_i|^2 = \frac{1}{2}\Big[\underbrace{|u_{11}|^2 + |u_{12}|^2}_{1} + \underbrace{|u_{21}|^2 + |u_{22}|^2}_{1}\Big] = 1.$$

Note. The result of this exercise demonstrates that the vector (a_0, a_1, a_2, a_3) of the Pauli basis expansion coefficients of the unitary matrix is normalized to unity.

10. *Answer.* $\beta^2 = I$, where I is the identity matrix of size 4×4.

11. *Answer.*

$$\alpha_1\alpha_2\alpha_3\beta = \begin{bmatrix} 0 & 0 & -i & 0 \\ 0 & 0 & 0 & -i \\ i & 0 & 0 & 0 \\ 0 & i & 0 & 0 \end{bmatrix}.$$

12. *Solution.*

Let us use the definition (2.14):

$$\sigma_1 \otimes \sigma_3 = \begin{bmatrix} 0 & 1 \\ 1 & 0 \end{bmatrix} \otimes \begin{bmatrix} 1 & 0 \\ 0 & -1 \end{bmatrix} = \begin{bmatrix} 0 \times \sigma_3 & 1 \times \sigma_3 \\ 1 \times \sigma_3 & 0 \times \sigma_3 \end{bmatrix} = \begin{bmatrix} 0 & 0 & 1 & 0 \\ 0 & 0 & 0 & -1 \\ 1 & 0 & 0 & 0 \\ 0 & -1 & 0 & 0 \end{bmatrix}.$$

In the same way, we obtain the remaining tensor products:

$$\sigma_3 \otimes \sigma_1 = \begin{bmatrix} 0 & 1 & 0 & 0 \\ 1 & 0 & 0 & 0 \\ 0 & 0 & 0 & -1 \\ 0 & 0 & -1 & 0 \end{bmatrix}, \quad \sigma_2 \otimes \sigma_3 = \begin{bmatrix} 0 & 0 & -i & 0 \\ 0 & 0 & 0 & i \\ i & 0 & 0 & 0 \\ 0 & -i & 0 & 0 \end{bmatrix},$$

$$\sigma_3 \otimes \sigma_2 = \begin{bmatrix} 0 & -i & 0 & 0 \\ i & 0 & 0 & 0 \\ 0 & 0 & 0 & i \\ 0 & 0 & -i & 0 \end{bmatrix}.$$

13. *Answer.*

Using the matrix representation $|0\rangle = \begin{bmatrix} 1 \\ 0 \end{bmatrix}$ and $|1\rangle = \begin{bmatrix} 0 \\ 1 \end{bmatrix}$, we obtain

$$\sigma_1 = |0\rangle \otimes \langle 1| + |1\rangle \otimes \langle 0|,$$
$$\sigma_2 = -i\,|0\rangle \otimes \langle 1| + i\,|1\rangle \otimes \langle 0|,$$
$$\sigma_3 = |0\rangle \otimes \langle 0| - |1\rangle \otimes \langle 1|.$$

Note that another matrix from the Pauli basis (the identity matrix of size 2×2) can be represented as $I = |0\rangle \otimes \langle 0| + |1\rangle \otimes \langle 1|$.

14. *Answer.* Hermitian matrices are (1), (2), (7), and (8).

15. *Answer.* Matrices (2), (4), and (8) are unitary.

16. *Answer.* Matrix M belongs to the unitary matrix class, but it is not Hermitian.

17. *Proof.*

Taking into account that $D_N^\dagger = D_N$, let us calculate the product $D_N D_N^\dagger$:

$$D_N D_N^\dagger = \begin{bmatrix} -1 + \dfrac{2}{N} & \dfrac{2}{N} & \cdots & \dfrac{2}{N} \\ \dfrac{2}{N} & -1 + \dfrac{2}{N} & \cdots & \dfrac{2}{N} \\ & \cdots\cdots\cdots\cdots & & \\ \dfrac{2}{N} & \dfrac{2}{N} & \cdots & -1 + \dfrac{2}{N} \end{bmatrix}$$

$$\times \begin{bmatrix} -1 + \dfrac{2}{N} & \dfrac{2}{N} & \cdots & \dfrac{2}{N} \\ \dfrac{2}{N} & -1 + \dfrac{2}{N} & \cdots & \dfrac{2}{N} \\ & \cdots\cdots\cdots\cdots & & \\ \dfrac{2}{N} & \dfrac{2}{N} & \cdots & -1 + \dfrac{2}{N} \end{bmatrix}.$$

The diagonal elements of the product are equal to

$$\left(D_N D_N^\dagger\right)_{ii} = \left(-1+\frac{2}{N}\right)^2 + \underbrace{\frac{2}{N}\cdot\frac{2}{N} + \frac{2}{N}\cdot\frac{2}{N} + \cdots + \frac{2}{N}\cdot\frac{2}{N}}_{(N-1)\text{ times}}$$

$$= \left(-1+\frac{2}{N}\right)^2 + (N-1)\frac{4}{N^2} = 1$$

for $i = 1, 2, \ldots, N$. In turn, non-diagonal matrix elements of $D_N D_N^\dagger$ are equal to

$$\left(D_N D_N^\dagger\right)_{ij} = \left(-1+\frac{2}{N}\right)\frac{2}{N} + \frac{2}{N}\left(-1+\frac{2}{N}\right)$$

$$+ \underbrace{\frac{2}{N}\cdot\frac{2}{N} + \frac{2}{N}\cdot\frac{2}{N} + \cdots + \frac{2}{N}\cdot\frac{2}{N}}_{(N-2)\text{ times}}$$

$$= \frac{4}{N}\left(-1+\frac{2}{N}\right) + (N-2)\frac{4}{N^2} = 0$$

for $i, j = 1, 2, \ldots, N$ and $i \neq j$.
Thus,

$$D_N D_N^\dagger = \begin{bmatrix} 1 & 0 & \ldots & 0 \\ 0 & 1 & \ldots & 0 \\ & & \ldots & \\ 0 & 0 & \ldots & 1 \end{bmatrix}.$$

The identity matrix is obtained. Hence, D_N is a unitary matrix for all natural N.
Note. Matrix P that meets the property $P^2 = P$ is called *projection matrix*, or *projector*. It is easy to see that D_N is a projector. It is known that eigenvalues of any projection matrix only take values 0 and 1.

18. *Answer.*
All such matrices have the following form

$$\begin{bmatrix} a & b \\ b & a \end{bmatrix},$$

where $a, b \in \mathbb{R}$.

19. *Solution.*
An arbitrary matrix M of size 2×2 contains exactly four complex parameters m_{ij}, where $i, j = 1, 2$:

$$M = \begin{bmatrix} m_{11} & m_{12} \\ m_{21} & m_{22} \end{bmatrix}.$$

Let us write down the condition that the commutator of matrices M and $\begin{bmatrix} 1 & 1 \\ 1 & 1 \end{bmatrix}$
is equal to zero matrix:

$$\begin{bmatrix} m_{11} & m_{12} \\ m_{21} & m_{22} \end{bmatrix} \times \begin{bmatrix} 1 & 1 \\ 1 & 1 \end{bmatrix} - \begin{bmatrix} 1 & 1 \\ 1 & 1 \end{bmatrix}\begin{bmatrix} m_{11} & m_{12} \\ m_{21} & m_{22} \end{bmatrix} = O,$$

where O is the zero matrix of the second order, i.e., it has the size 2×2. Thus, we come to a linear system of equations relatively unknown m_{ij}:

$$\begin{cases} (m_{11} + m_{12}) - (m_{11} + m_{21}) = 0, \\ (m_{11} + m_{12}) - (m_{12} + m_{22}) = 0, \\ (m_{21} + m_{22}) - (m_{11} + m_{21}) = 0, \\ (m_{21} + m_{22}) - (m_{12} + m_{22}) = 0, \end{cases}$$

or, after simplifications,

$$\begin{cases} m_{12} - m_{21} = 0, \\ m_{11} - m_{22} = 0. \end{cases}$$

Consequently, any matrix commuting with $\begin{bmatrix} 1 & 1 \\ 1 & 1 \end{bmatrix}$, has the form $M = \begin{bmatrix} m_{11} & m_{12} \\ m_{12} & m_{11} \end{bmatrix}$, where m_{11} and m_{12} are some complex numbers.

Next, let us use the unitarity condition $M M^\dagger = I$, where I is the identity matrix:

$$\begin{bmatrix} m_{11} & m_{12} \\ m_{12} & m_{11} \end{bmatrix} \times \begin{bmatrix} m_{11} & m_{12} \\ m_{12} & m_{11} \end{bmatrix}^\dagger = \begin{bmatrix} m_{11} & m_{12} \\ m_{12} & m_{11} \end{bmatrix} \times \begin{bmatrix} m_{11}^* & m_{12}^* \\ m_{12}^* & m_{11}^* \end{bmatrix} = \begin{bmatrix} 1 & 0 \\ 0 & 1 \end{bmatrix},$$

which results in a system of equations

$$\begin{cases} |m_{11}|^2 + |m_{12}|^2 = 1, \\ m_{11}m_{12}^* + m_{12}m_{11}^* = 0, \\ m_{12}m_{11}^* + m_{11}m_{12}^* = 0, \\ |m_{12}|^2 + |m_{11}|^2 = 1. \end{cases}$$

We leave only independent equations in the obtained system.

$$\begin{cases} |m_{11}|^2 + |m_{12}|^2 = 1, \\ m_{11}m_{12}^* + m_{12}m_{11}^* = 0. \end{cases}$$

Let us proceed to the trigonometric representation of complex numbers $m_{11} = \rho_1 e^{i\varphi_1}$, $m_{12} = \rho_2 e^{i\varphi_2}$, where $\rho_1, \rho_2 \geqslant 0$, $\varphi_1, \varphi_2 \in [0, 2\pi)$. Conditions for $\rho_{1,2}$ and $\varphi_{1,2}$ take the following form:

$$\begin{cases} |\rho_1|^2 + |\rho_2|^2 = 1, \\ \rho_1\rho_2\left(e^{i(\varphi_1-\varphi_2)} + e^{-i(\varphi_1-\varphi_2)}\right) = 0. \end{cases}$$

These conditions are satisfied by the following parameter values:
(1) $\rho_1 = 0$, $\rho_2 = e^{i\delta}$, and $\delta \in \mathbb{R}$,
(2) $\rho_2 = 0$, $\rho_1 = e^{i\delta'}$, and $\delta' \in \mathbb{R}$,
(3) $\rho_1\rho_2 \geqslant 0$, $\cos(\varphi_1 - \varphi_2) = 0 \Leftrightarrow \varphi_2 = \varphi_1 + \dfrac{\pi}{2} + \pi k, k \in \mathbb{Z}$.

As a result, matrix M can be represented in the following form:

$$M = \begin{bmatrix} \rho_1 e^{i\varphi_1} & \pm i\sqrt{1 - \rho_1^2}\, e^{i\varphi_1} \\ \pm i\sqrt{1 - \rho_1^2}\, e^{i\varphi_1} & \rho_1 e^{i\varphi_1} \end{bmatrix} = e^{i\varphi_1} \begin{bmatrix} \rho_1 & \pm i\sqrt{1 - \rho_1^2} \\ \pm i\sqrt{1 - \rho_1^2} & \rho_1 \end{bmatrix},$$

where $\rho_1 \geqslant 0$ and $\varphi_1 \in [0, 2\pi)$.

Using the notations $\rho_1 = \cos \chi, \chi \in [0, 2\pi)$, and taking into account the relation $\pm\sqrt{1 - \rho_1^2} = \sin \chi$, we obtain:

$$M = \begin{bmatrix} \cos \chi & i \sin \chi \\ i \sin \chi & \cos \chi \end{bmatrix}. \tag{10.40}$$

Any matrix that commutates with the real matrix $\begin{bmatrix} 1 & 1 \\ 1 & 1 \end{bmatrix}$ has the form (10.40).

20. *Solution.*

Let us prove that a unitary matrix V, such that the properties $V^2 = U$ and $VV^\dagger = I$ are fulfilled, exists for any unitary matrix U of the second order. To do this, convert U with the unitary transform A to a diagonal form:

$$U = A \begin{bmatrix} \lambda_1 & 0 \\ 0 & \lambda_2 \end{bmatrix} A^\dagger,$$

where $\lambda_{1,2}$ are eigenvalues of matrix U. As it is known, one of the features of unitary matrices is the equality to one of modules of eigenvalues (see Appendix B on page 109): $|\lambda_{1,2}| = 1$. This allows them to be represented as exponents with an imaginary argument $\lambda_{1,2} = e^{i\varphi_{1,2}}$, where $\varphi_{1,2} \in \mathbb{R}$.

Next, consider the matrix

$$V = A \times \begin{bmatrix} \sqrt{\lambda_1} & 0 \\ 0 & \sqrt{\lambda_2} \end{bmatrix} \times A^\dagger = A \times \begin{bmatrix} e^{i\varphi_1/2} & 0 \\ 0 & e^{i\varphi_2/2} \end{bmatrix} \times A^\dagger,$$

which uses the principal value of the complex square root.

It is clear that the matrix V built in this way is unitary and has the property $V^2 = U$.

21. *Solution.*

Qubit is described by a vector in a two-dimensional Hilbert space

$$|\psi\rangle = u |0\rangle + v |1\rangle ,$$

where u, v are some complex numbers. The parameters $u = \text{Re}\, u + \text{Im}\, u$ and $v = \text{Re}\, v + \text{Im}\, u$ are subject to normalization condition

$$|u|^2 + |v|^2 = 1.$$

Consequently, of the four real parameters $\text{Re}\, u$, $\text{Im}\, u$, $\text{Re}\, v$ and $\text{Im}\, u$ only three are independent.

As a result, three real parameters are necessary and sufficient to set an arbitrary state of one qubit. Note that if we consider that the states of qubits which differ only by the global phase factor are identical: $|\psi'\rangle = e^{i\delta} |\psi\rangle$, the specified number of real parameters will be equal to two.

Thus, it is possible to parameterize an arbitrary state of a qubit in the form

$$|\psi\rangle = \cos(\theta/2) |0\rangle + e^{i\varphi} \sin(\theta/2) |1\rangle ,$$

where $\theta \in [0, \pi)$ and $\varphi \in [0, 2\pi)$ are real numbers.

22. *Solution.*

(1) The properties of the fidelity are derived from the orthonormalization properties of complex vectors. In fact, if the states $|\psi_1\rangle$ and $|\psi_2\rangle$ coincide, i.e., the equalities $|\psi_1\rangle = |\psi_2\rangle = |\psi\rangle$ are valid, then

$$\mathcal{F} = |\langle\psi_1|\psi_1\rangle|^2 = 1^2 = 1.$$

If the states $|\psi_1\rangle$ and $|\psi_2\rangle$ are orthogonal, we obtain

$$\mathcal{F} = |\langle\psi_1|\psi_2\rangle|^2 = 0$$

by virtue of equality to zero of scalar product of orthogonal vectors.

Note. They say that the value $\mathcal{F}$ characterizes closeness of states in the vector space of states. The value $\mathcal{F} = 1$ reflects the state identity, while $\mathcal{F} = 0$ reflects the maximum "noncoincidence" of such states [3,4].

(2) Let us use the definition of the value $\mathcal{F}$.

$$\mathcal{F} = |\langle\psi_1|\psi_2\rangle|^2$$
$$= \frac{1}{4}\left|[|0\rangle + e^{i\varphi_1}|1\rangle]^*[|0\rangle + e^{i\varphi_2}|1\rangle]\right|^2$$
$$= \frac{1}{4}\left|[\langle0| + e^{-i\varphi_1}\langle1|][|0\rangle + e^{i\varphi_2}|1\rangle]\right|^2$$
$$= \frac{1}{\sqrt{4}}\left|\langle0|0\rangle + e^{i\varphi_2}\langle0|1\rangle + e^{-i\varphi_1}\langle1|0\rangle + e^{i(\varphi_2-\varphi_1)}\langle1|1\rangle\right|^2.$$

By virtue of the orthonormality of $\langle i|j\rangle = \delta_{ij}$, we obtain

$$\mathcal{F} = \frac{1}{4}|1 + e^{i(\varphi_2-\varphi_1)}|^2$$
$$= \frac{1}{4}[2 + 2\cos(\varphi_2 - \varphi_1)] = \cos^2[(\varphi_2 - \varphi_1)/2].$$

23. *Solution.*

According to the definition of the fidelity, we obtain

$$\mathcal{F} = |\langle\psi_1|\psi_2\rangle|^2$$
$$= |\{e^{i\gamma_1}[\cos(\theta_1/2)|0\rangle + e^{i\varphi_1}\sin(\theta_1/2)|1\rangle]\}^*$$
$$\times\{e^{i\gamma_2}[\cos(\theta_2/2)|0\rangle + e^{i\varphi_2}\sin(\theta_2/2)|1\rangle]\}|^2$$
$$= |e^{i(\gamma_2-\gamma_1)}[\cos(\theta_1/2)\langle0| - e^{i\varphi_1}\sin(\theta_1/2)\langle1|]$$
$$\times[\cos(\theta_2/2)\langle0| + e^{i\varphi_2}\sin(\theta_2/2)|1\rangle]|^2$$
$$= |e^{i(\gamma_2-\gamma_1)}|^2|\cos(\theta_1/2)\cos(\theta_2/2)\langle0|0\rangle + \cos(\theta_1/2)e^{i\varphi_2}\sin(\theta_2/2)\langle0|1\rangle$$
$$- \cos(\theta_2/2)e^{i\varphi_1}\sin(\theta_1/2)\langle1|0\rangle - e^{i(\varphi_2-\varphi_1)}\sin(\theta_1/2)\sin(\theta_2/2)\langle1|1\rangle|^2.$$

Let us perform algebraic transformations considering the orthonormality condition $\langle i|j \rangle = 1$, if $i = j$, and $\langle i|j \rangle = 0$ if $i \neq j$.

$$\mathcal{F} = \underbrace{|e^{i(\gamma_2 - \gamma_1)}|^2}_{1} |\cos(\theta_1/2)\cos(\theta_2/2) - e^{i(\varphi_2 - \varphi_1)}\sin(\theta_1/2)\sin(\theta_2/2)|^2$$

$$= [\cos(\theta_1/2)\cos(\theta_2/2) - e^{i(\varphi_2 - \varphi_1)}\sin(\theta_1/2)\sin(\theta_2/2)]$$
$$\times [\cos(\theta_1/2)\cos(\theta_2/2) + e^{i(\varphi_2 - \varphi_1)}\sin(\theta_1/2)\sin(\theta_2/2)]$$
$$= [\cos(\theta_1/2)\cos(\theta_2/2)]^2 + [\sin(\theta_1/2)\sin(\theta_2/2)]^2$$
$$+ 2\cos(\theta_1/2)\cos(\theta_2/2)\sin(\theta_1/2)\sin(\theta_2/2)\cos(\varphi_2 - \varphi_1).$$

Then we will consider the trigonometric formulas

$$\cos^2 \frac{a}{2} = \frac{1}{2}(1 + \cos 2a),$$
$$\sin^2 \frac{a}{2} = \frac{1}{2}(1 - \cos 2a),$$
$$\sin 2a = 2\sin a \cos a,$$

valid for all real a. With these equalities taken into account, we will obtain

$$\mathcal{F} = \frac{1}{2}[1 + \cos\theta_1 \cos\theta_2 + \sin\theta_1 \sin\theta_2 \cos(\varphi_2 - \varphi_1)].$$

For the purpose of further simplification, consider two unity vectors in the space $\mathbb{R}^3$

$$\boldsymbol{n}_1 = \begin{bmatrix} \sin\theta_1 \cos\varphi_1 \\ \sin\theta_1 \sin\varphi_1 \\ \cos\theta_1 \end{bmatrix}, \quad \boldsymbol{n}_2 = \begin{bmatrix} \sin\theta_2 \cos\varphi_2 \\ \sin\theta_2 \sin\varphi_2 \\ \cos\theta_2 \end{bmatrix}.$$

The vectors $\boldsymbol{n}_1$ and $\boldsymbol{n}_2$ clearly represent the state of qubits of $|\psi_1\rangle$ and $|\psi_2\rangle$ on the unity sphere. With the help of a scalar product, it is easy to calculate the cosine of the angle Δ between the specified vectors:

$$\cos\Delta = \cos\theta_1 \cos\theta_2 + \sin\theta_1 \sin\theta_2 \cos(\varphi_2 - \varphi_1).$$

Comparing $\cos\Delta$ with the expression for $\mathcal{F}$, we conclude:

$$\mathcal{F} = \frac{1}{2}[1 + \cos\Delta],$$

or

$$\mathcal{F} = \cos^2(\Delta/2),$$

where Δ is the angle between vectors $\boldsymbol{n}_1$ and $\boldsymbol{n}_2$ in real space. In this form of answer recording, we clearly see the fidelity properties: $0 \leqslant \mathcal{F} \leqslant 1$, and $\mathcal{F} = 1$ for matching quantum states.

24. *Proof.*

Let us introduce the vector $|z\rangle = |x\rangle - \lambda|y\rangle$, where λ is a complex number. Based on the first property of a scalar product (see page 3) write down $\langle z|z\rangle \geqslant 0$, or

$$\langle z|z\rangle = ((\langle x| - \lambda^*\langle y|)(|x\rangle - \lambda|y\rangle))$$
$$= \langle x|x\rangle - \lambda\langle x|y\rangle - \lambda^*\langle y|x\rangle + |\lambda|^2\langle y|y\rangle \geqslant 0.$$

This inequality must be valid for any $\lambda \in \mathbb{C}$.

Next, consider two cases: $|y\rangle = 0$ and $|y\rangle \neq 0$.

1. Let $|y\rangle = 0$. Then Schwarz inequality turns into an identity $0 = 0$.
2. Let $|y\rangle \neq 0$. Suppose $\lambda = \dfrac{\langle y|x\rangle}{\langle y|y\rangle}$:

$$\langle z|z\rangle = \langle x|x\rangle - \frac{|\langle x|y\rangle|^2}{\langle y|y\rangle} \geqslant 0.$$

The resulting inequality can be rewritten in the following form:

$$|\langle x|y\rangle|^2 \leqslant \langle x|x\rangle\langle y|y\rangle.$$

Thus, the Schwarz inequality is proved.

Inequality (10.29) goes into equality if $|y\rangle = a|x\rangle$ for some $a \in \mathbb{C}$. In reverse, if the equality $|\langle x|y\rangle|^2 = \langle x|x\rangle\langle y|y\rangle$ is valid, this entails $\langle z|z\rangle = 0$ and, in turn, $|x\rangle = \lambda|y\rangle$, where $\lambda \in \mathbb{C}$. Consequently, equality in (10.29) is met if and only if vectors $|x\rangle$ and $|y\rangle$ differ by a multiplier, i.e., the vectors are proportional.

25. *Proof.*

Let us use the properties of the scalar product (see page 3):

$$\langle \psi_1 + \psi_2|\psi_1 + \psi_2\rangle = \langle \psi_1|\psi_1\rangle + \langle \psi_1|\psi_2\rangle + \langle \psi_2|\psi_1\rangle + \langle \psi_2|\psi_2\rangle.$$

The summands $\langle \psi_1|\psi_2\rangle$ and $\langle \psi_2|\psi_1\rangle$ turn to zero because of the orthogonality of $|\psi_1\rangle$ and $|\psi_2\rangle$. This way, for all $|\psi_1\rangle$ and $|\psi_2\rangle$, the equality is valid $\langle \psi_1 + \psi_2|\psi_1 + \psi_2\rangle = \langle \psi_1|\psi_1\rangle + \langle \psi_2|\psi_2\rangle$. Pythagoras theorem is proved.

26. *Answer.*

$$(1)\ |\psi\rangle = \begin{bmatrix} 0 \\ 0 \\ 1 \\ 0 \\ 0 \\ 0 \\ 0 \\ 0 \end{bmatrix};$$

(2) $|\psi\rangle = \begin{bmatrix} 0 \\ 0 \\ 1 \\ 0 \\ 0 \\ 1 \\ 0 \\ 0 \end{bmatrix}$.

27. *Answer.*

(1) $|00\ldots011\rangle = \begin{bmatrix} 0 \\ 0 \\ 0 \\ 1 \\ 0 \\ \vdots \\ 0 \\ 0 \end{bmatrix}$;

(2) $|11\ldots10\rangle = \begin{bmatrix} 0 \\ 0 \\ \vdots \\ 0 \\ 1 \\ 0 \end{bmatrix}$.

28. *Answer.*

$$|\psi\rangle = \frac{1}{10}(|000\rangle + 2\,|001\rangle - 4\,|010\rangle + i\,|011\rangle$$
$$+ i\,|100\rangle - |101\rangle + |110\rangle + 5\sqrt{3}\,|111\rangle).$$

29. *Solution.*
We perform the multiplication operation by removing parentheses:

$$|\psi\rangle\,|\psi\rangle = \frac{1}{\sqrt{2}}(|0\rangle - i\,|1\rangle)\frac{1}{\sqrt{2}}(|0\rangle - i\,|1\rangle) = \frac{1}{2}(|0\rangle\,|0\rangle - i\,|0\rangle\,|1\rangle - i\,|1\rangle\,|0\rangle - |1\rangle\,|1\rangle).$$

Let us write down the final answer in the following form:

$$|\psi\rangle\,|\psi\rangle = \frac{1}{2}(|00\rangle - i\,|01\rangle - i\,|10\rangle - |11\rangle).$$

30. *Solution.*
We introduce designations for basis elements:

$$|e_1\rangle = \frac{1}{10}\begin{bmatrix} 1 \\ 3i \end{bmatrix}, \quad |e_2\rangle = \frac{1}{10}\begin{bmatrix} 3i \\ 1 \end{bmatrix}.$$

Let us check the compliance with the conditions $\langle e_i|e_j\rangle = \delta_{ij}$ for $i,j = 1,2$:

$$\langle e_1|e_1\rangle = \frac{1}{10}\begin{bmatrix} 1 \\ 3i \end{bmatrix}^{\dagger} \times \frac{1}{10}\begin{bmatrix} 1 \\ 3i \end{bmatrix} = \frac{1}{10}[1 \ -3i] \times \frac{1}{10}\begin{bmatrix} 1 \\ 3i \end{bmatrix} = \frac{1}{10}(1 + (-3i)3i) = 1,$$

$$\langle e_1 | e_2 \rangle = \frac{1}{10}\begin{bmatrix} 1 \\ 3i \end{bmatrix}^{\dagger} \times \frac{1}{10}\begin{bmatrix} 3i \\ 1 \end{bmatrix} = \frac{1}{10}[1 \ -3i] \times \frac{1}{10}\begin{bmatrix} 3i \\ 1 \end{bmatrix} = 0;$$

$$\langle e_2 | e_1 \rangle = \frac{1}{10}\begin{bmatrix} 3i \\ 1 \end{bmatrix}^{\dagger} \times \frac{1}{10}\begin{bmatrix} 1 \\ 3i \end{bmatrix} = \frac{1}{10}[-3i \ 1] \times \frac{1}{10}\begin{bmatrix} 1 \\ 3i \end{bmatrix} = 0,$$

$$\langle e_2 | e_2 \rangle = \frac{1}{10}\begin{bmatrix} 3i \\ 1 \end{bmatrix}^{\dagger} \times \frac{1}{10}\begin{bmatrix} 3i \\ 1 \end{bmatrix} = \frac{1}{10}[-3i \ 1] \times \frac{1}{10}\begin{bmatrix} 3i \\ 1 \end{bmatrix} = \frac{1}{10}\left((-3i)3i + 1\right) = 1.$$

Consequently, the system $\{|e_1\rangle , |e_2\rangle\}$ is orthonormalized.

31. *Solution.*

$$M = \begin{cases} 0 \ \text{with a probability } p_0 = \dfrac{1}{18}, \\[2mm] 1 \ \text{with a probability } p_1 = \dfrac{17}{18}. \end{cases}$$

32. *Solution.*

The wave function of qubit $|\psi\rangle = u\,|0\rangle + v\,|1\rangle$ has the following coefficients in the superposition of basis states: $u = \dfrac{1 + 2i}{5}, v = \dfrac{2 + 4i}{5}$.

The result of the quantum measurement M in the basis $\{|0\rangle , |1\rangle\}$ is zero or one with the probabilities $|u|^2$ and $|v|^2$, respectively:

$$M = \begin{cases} 0 \ \text{with a probability } p_0 = \left(\dfrac{5+i}{6}\right)\left(\dfrac{5+i}{6}\right)^{*} = \dfrac{13}{18}, \\[3mm] 1 \ \text{with a probability } p_1 = \left(\dfrac{1-3i}{6}\right)\left(\dfrac{1-3i}{6}\right)^{*} = \dfrac{5}{18}. \end{cases}$$

Consider the measurement procedure in the basis $B_2 = \{|+\rangle , |-\rangle\}$. We express the vectors of the computational basis through $|+\rangle$ and $|-\rangle\}$:

$$|0\rangle = \frac{1}{\sqrt{2}}(|+\rangle + |-\rangle),$$

$$|1\rangle = \frac{1}{\sqrt{2}}(|+\rangle + |-\rangle).$$

Consequently,

$$\begin{aligned} |\psi\rangle &= \frac{5+i}{6}|0\rangle + \frac{1-3i}{6}|1\rangle \\ &= \frac{5+i}{6\sqrt{2}}(|+\rangle + |-\rangle) + \frac{1-3i}{6\sqrt{2}}(|+\rangle + |-\rangle) \\ &= \frac{6-2i}{6\sqrt{2}}|+\rangle + \frac{4+4i}{6\sqrt{2}}|-\rangle \\ &= \frac{3-i}{3\sqrt{2}}|+\rangle + 2 \times \frac{1+i}{3\sqrt{2}}|-\rangle. \end{aligned}$$

The probability of finding the system in the state $|+\rangle$ is equal to

$$p_+ = \left(\frac{3-i}{3\sqrt{2}}\right) \times \left(\frac{3-i}{3\sqrt{2}}\right)^* = \frac{5}{9}.$$

The same way we obtain the probability to detect the system in the state $|-\rangle$:

$$p_- = \left(2\frac{1+i}{3\sqrt{2}}\right) \times \left(2\frac{1+i}{3\sqrt{2}}\right)^* = \frac{4}{9}.$$

The equality $p_+ + p_- = 1$ is valid, reflecting the law of total probability preservation.

35. *Solution.*

The state of the qubit $|0\rangle$ is described by the matrix $\begin{bmatrix} 1 \\ 0 \end{bmatrix}$. Taking into account the matrix representation of quantum elements H, S, and X from Table 3.1 on page 13, we obtain

$$|\psi\rangle \rightarrow |\psi'\rangle$$

$$= \frac{1}{\sqrt{2}}\begin{bmatrix} 1 & 1 \\ 1 & -1 \end{bmatrix}\begin{bmatrix} 1 & 0 \\ 0 & i \end{bmatrix} \times \frac{1}{\sqrt{2}}\begin{bmatrix} 1 & 1 \\ 1 & -1 \end{bmatrix}\begin{bmatrix} 0 & 1 \\ 1 & 0 \end{bmatrix}\begin{bmatrix} 1 \\ 0 \end{bmatrix} = \begin{bmatrix} \dfrac{1-i}{2} \\ \dfrac{1+i}{2} \end{bmatrix}.$$

Thus, as a result of the action of the quantum circuit on a qubit, which is in the state $|\psi\rangle = |0\rangle$, it will convert to the state

$$|\psi'\rangle = \frac{1-i}{2}\,|0\rangle + \frac{1+i}{2}\,|1\rangle .$$

36. *Answer.* As a result of the action of the quantum circuit on a qubit, which is in the state $|\psi\rangle = |1\rangle$, it will convert to the state

$$|\psi'\rangle = \left(-\frac{i}{\sqrt{2}}\right)|0\rangle - \left(\frac{1-i}{2}\right)|1\rangle .$$

37. *Answer.* The qubit, which is initially in the state $|\psi\rangle = u\,|0\rangle + v\,|1\rangle$, will convert to the state $|\psi'\rangle = \left(\frac{1}{2}(1+i)u + \frac{1}{\sqrt{2}}v\right)|0\rangle + \left(\frac{1}{2}(1+i)u - \frac{1}{\sqrt{2}}v\right)|1\rangle$.

39. *Solution.*

The quantum state $|\Psi\rangle$ of the system after the action of the Hadamard operator will be described by the vector $|\Psi'\rangle$:

$$|\Psi\rangle \rightarrow |\Psi'\rangle = \frac{1}{2}\big((|0\rangle + |1\rangle)\,|\psi_1\rangle + (|0\rangle - |1\rangle)\,|\psi_2\rangle\big).$$

Grouping similar terms, we conclude that

$$|\Psi'\rangle = a_0\,|0\rangle + a_1\,|1\rangle ,$$

where coefficients $a_0 = \frac{1}{2}(|\psi_1\rangle + |\psi_2\rangle)$ and $a_1 = \frac{1}{2}(|\psi_1\rangle - |\psi_2\rangle)$ are the probability amplitudes corresponding to the state of the first qubit $|0\rangle$ and $|1\rangle$, respectively. Next, we calculate the square of the modulus a_0.

$$|a_0|^2 = a_0 a_0^* = \frac{1}{2}(\langle\psi_1| + \langle\psi_2|) \times \frac{1}{2}(|\psi_1\rangle + |\psi_2\rangle)$$

$$= \frac{1}{4}(\langle\psi_1|\psi_1\rangle + \langle\psi_1|\psi_2\rangle + \langle\psi_2|\psi_1\rangle + \langle\psi_2|\psi_2\rangle).$$

Taking into account the normalization of vectors $|\psi_1\rangle$ and $|\psi_2\rangle$, and the symmetry properties of $\langle\psi_1|\psi_2\rangle = \langle\psi_2|\psi_1\rangle^*$ (see page 3), we obtain

$$|a_0|^2 = \frac{1}{4}(1 + \langle\psi_1|\psi_2\rangle + \langle\psi_2|\psi_1\rangle + 1) = \frac{1}{2}(1 + \text{Re}\,\langle\psi_1|\psi_2\rangle).$$

Similar calculations lead to equality

$$|a_1|^2 = \frac{1}{2}(1 - \text{Re}\,\langle\psi_1|\psi_2\rangle).$$

As a result, the result of measuring M of the first qubit is equal to

$$M = \begin{cases} 0 \text{ with probability } p_0 = \frac{1}{2}(1 + \text{Re}\,\langle\psi_1|\psi_2\rangle), \\[2mm] 1 \text{ with probability } p_1 = \frac{1}{2}(1 - \text{Re}\,\langle\psi_1|\psi_2\rangle). \end{cases}$$

40. *Solution.*

After applying the CNOT operation, we obtain the following:

$$|\psi\rangle \to |\psi'\rangle = \sqrt{\frac{3}{5}}|11\rangle + i\sqrt{\frac{2}{5}}|10\rangle.$$

Further, Hadamard operation with the second qubit will bring the system to the state $|\psi''\rangle$:

$$|\psi'\rangle \to |\psi''\rangle = \sqrt{\frac{3}{5}}|1\rangle(|0\rangle - |1\rangle)/\sqrt{2} + i\sqrt{\frac{2}{5}}|1\rangle(|0\rangle + |1\rangle)/\sqrt{2}$$

$$= \frac{\sqrt{3} + i\sqrt{2}}{\sqrt{10}}|10\rangle - \frac{\sqrt{3} - i\sqrt{2}}{\sqrt{10}}|11\rangle.$$

41. *Answer.* $U_H |q\rangle = \frac{1}{\sqrt{2}}(|0\rangle + (-1)^q |1\rangle).$

42. *Answer.* $|\psi_{\text{out}}\rangle = u|01\rangle + v|10\rangle.$

45. *Answer.*

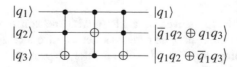

46. *Solution.*

Each of the Boolean functions that form the set $\{\vee, \wedge, 1\}$, has a monotone property. Therefore, it is impossible to obtain constant 0 by conjunction, disjunction, and constant 1. Consequently, the set $\{\vee, \wedge, 1\}$ is not universal.

47. *Solution.*

It is impossible to build a Toffoli gate using only Fredkin gates, without auxiliary qubits. It follows from the property of the Fredkin gate to save the number of zeros and ones.

Obviously, since Fredkin gate is universal, any quantum circuit, including CC-NOT, can be implemented if there are auxiliary qubits. One way is to use three instances of Fredkin gate if you have three auxiliary qubits, two of which are pre-installed in $|0\rangle$, the third is pre-installed in $|1\rangle$. The quantum circuit obtained in this way is shown below:

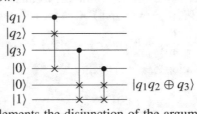

48. *Answer.* The circuit implements the disjunction of the arguments $f(q_1, q_2) = q_1 \vee q_2$.

49. *Solution.*

As shown in Example 6.2 on page 33, operation $\underbrace{CC...C}_{n \text{ times}} NOT$ can be built from

two symmetrically arranged circuits for $\underbrace{CC...C}_{n-1 \text{ times}} NOT$ and one Toffoli gate placed

in the center. The CCCNOT circuit contains three Toffoli gates. Therefore, for the full number of gates, we get a recurrence relation:

$$\begin{cases} R_n = 2R_{n-1} + 1, \\ R_3 = 3. \end{cases}$$

Its solution, as easy to see, is the function

$$R_n = 2^{n-1} - 1 \text{ for } n \geqslant 3.$$

Therefore, this circuit contains an exponential number of elementary gates.

50. *Answer.*

$$U_{CCNOT} |\psi\rangle = \frac{1}{2}(|000\rangle + |001\rangle + |100\rangle - |110\rangle),$$

$$U_{CSWAP} |\psi\rangle = \frac{1}{2}(|000\rangle + |001\rangle + |100\rangle - |111\rangle) = |\psi\rangle.$$

51. *Solution.*

The problem is solved by analogy with the construction of Toffoli gate from two-qubit elements (see page 34). Let us introduce a unitary operator V, which has the property $V^2 = U$. This operator, as proven in Exercise 20, exists for any U and is unitary.

Then the next quantum circuit implements the controlled unitary operation U:

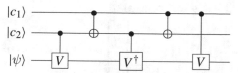

To prove the correctness of the built quantum circuit, let us consider several cases.

(1) $|c_1\rangle = |0\rangle$. On the third qubit, there acts either an identical operator I or a product of operators $VV^\dagger$. Since $VV^\dagger \equiv I$, the third qubit does not change.

(2) $|c_1\rangle = |1\rangle$, $|c_2\rangle = |0\rangle$. The operator $V^\dagger V$ acts on the third qubit. The state of the third qubit does not change.

(3) $|c_1\rangle = |c_2\rangle = |1\rangle$. The third qubit $|q\rangle$ changes to $VV|q\rangle \equiv U|q\rangle$.

Thus, there is executed a unitary operation U controlled by two qubits $|c_1\rangle$ and $|c_2\rangle$.

52. *Answer.*

The student's statement is false because the CNOT operator generally does not leave the controlling qubit unchanged. The calculation using the matrix representation of the Hadamard operators U_H and the controlled NOT U_{CNOT} shows that the output state is $\dfrac{1}{2}(|00\rangle + |01\rangle - |10\rangle + |11\rangle)$.

53. *Answer.*

The system state will not change: $|\psi\rangle = (|00\rangle + |11\rangle)/\sqrt{2}$.

54. *Answer.*

The system state will not change: $|\psi\rangle = (|00\rangle + |11\rangle)/\sqrt{2}$.

55. *Answer.* $\alpha\delta = \beta\gamma$.

56. *Proof.*

Let us write down Bell state $|\beta_1\rangle$–$|\beta_4\rangle$ as a matrix notation:

$$|\beta_1\rangle = \frac{1}{\sqrt{2}}(|00\rangle + |11\rangle) = \frac{1}{\sqrt{2}}\begin{bmatrix} 1 \\ 0 \\ 0 \\ 1 \end{bmatrix},$$

$$|\beta_2\rangle = \frac{1}{\sqrt{2}}(|01\rangle + |10\rangle) = \frac{1}{\sqrt{2}}\begin{bmatrix} 0 \\ 1 \\ 1 \\ 0 \end{bmatrix},$$

$$|\beta_3\rangle = \frac{1}{\sqrt{2}}(|00\rangle - |11\rangle) = \frac{1}{\sqrt{2}}\begin{bmatrix} 1 \\ 0 \\ 0 \\ -1 \end{bmatrix},$$

$$|\beta_4\rangle = \frac{1}{\sqrt{2}}(|01\rangle - |10\rangle) = \frac{1}{\sqrt{2}}\begin{bmatrix} 0 \\ 1 \\ -1 \\ 0 \end{bmatrix}.$$

You can see that all states are normalized:

$$\langle\beta_1|\beta_1\rangle = \frac{1}{\sqrt{2}}[1\ 0\ 0\ 1] \times \frac{1}{\sqrt{2}}\begin{bmatrix}1\\0\\0\\1\end{bmatrix} = \frac{1}{2}(1^2 + 0^2 + 0^2 + 1^2) = 1,$$

analogous

$$\langle\beta_2|\beta_2\rangle = \langle\beta_3|\beta_3\rangle = \langle\beta_4|\beta_4\rangle = 1.$$

Let us check the orthogonality condition:

$$\forall i, j = 1, 2, 3, 4, \ i \neq j : \langle\beta_i|\beta_j\rangle = 0.$$

Direct calculation shows that

$$\langle\beta_1|\beta_2\rangle = \frac{1}{\sqrt{2}}[1\ 0\ 0\ 1] \times \frac{1}{\sqrt{2}}\begin{bmatrix}0\\1\\1\\0\end{bmatrix} = 0,$$

and further

$$\langle\beta_1|\beta_2\rangle = \langle\beta_1|\beta_3\rangle = \ldots = \langle\beta_2|\beta_1\rangle = \ldots = 0.$$

The system of vectors is orthogonal. Consequently, the Bell basis is an example of an orthonormalized basis.

57. *Solution.*
The student is wrong. His statement contradicts the theorem that it is impossible to copy an arbitrary state (see page 22). In fact the output of the quantum circuit will have the state $u\,|00\rangle + v\,|00\rangle$, which at $uv \neq 0$ does not match the intended copy of $(u\,|0\rangle + v\,|0\rangle)(u\,|0\rangle + v\,|0\rangle)$.

58. *Proof.*
Let us prove this using the method "by contradiction". Suppose that $|\text{GHZ}\rangle$ is not entangled, and prove that it leads to a contradiction.
In fact, with this assumption in mind, it is possible to record without restriction of the generality the following:

$$|\text{GHZ}\rangle = (a_1\,|0\rangle + a_2\,|1\rangle)(b_1\,|00\rangle + b_2\,|01\rangle + b_1\,|10\rangle + b_2\,|11\rangle),$$

where $a_1, a_2, b_1, \ldots, b_4$ are some complex numbers. Perform the multiplication operations and remove parentheses:

$$|\text{GHZ}\rangle = a_1 b_1\,|000\rangle + a_1 b_2\,|001\rangle + a_1 b_3\,|010\rangle + a_1 b_4\,|011\rangle$$
$$+ a_2 b_1\,|100\rangle + a_2 b_2\,|101\rangle + a_2 b_3\,|110\rangle + a_2 b_4\,|111\rangle.$$

Comparing the obtained sum with the expression $\dfrac{1}{\sqrt{2}}(|000\rangle + |111\rangle)$, we come to the system of equations concerning the coefficients $a_1, a_2, b_1, \ldots, b_4$:

$$\begin{cases} a_1 b_1 = \dfrac{1}{\sqrt{2}}, \\ a_1 b_2 = 0, \\ a_1 b_3 = 0, \\ a_1 b_4 = 0, \\ a_2 b_1 = 0, \\ a_2 b_2 = 0, \\ a_2 b_3 = 0, \\ a_2 b_4 = \dfrac{1}{\sqrt{2}}. \end{cases}$$

As it is easy to see, this system is incompatible. Consider, for example, the first, fifth and eighth equations:

$$a_1 b_1 = \frac{1}{\sqrt{2}}, \quad a_2 b_1 = 0, \quad a_2 b_4 = \frac{1}{\sqrt{2}}.$$

Since the product is $a_1 b_1 \neq 0$, and $a_2 b_1 = 0$, then $a_2 = 0$. This contradicts the equality $a_2 b_4 = \dfrac{1}{\sqrt{2}}$.

Consequently, the system of equations concerning the coefficients $a_1, a_2, b_1, \ldots, b_4$ has no solution. The state $|GHZ\rangle$ is entangled.

The entanglement of the state $|GHZ\rangle$ manifests itself in the fact that measuring any of the qubits will directly answer the question of what state all other qubits are in.

59. *Solution.*

The state $|GHZ\rangle$ is entangled (see Exercise 58). As a first step in the procedure of building $|GHZ\rangle$, we apply the Hadamard operation to the first qubit. After that he will change to the state $|\psi_1\rangle = (|0\rangle + |1\rangle)/\sqrt{2}$. Next, on the second step of building the GHZ-state to the second and third qubits, we apply the CNOT operation controlled by the first qubit $|\psi_1\rangle$.

Let us present the corresponding quantum circuit.

As it is known, as a result of the CNOT operation, the controlling and controlled qubits are added to modulo two and the result is saved in the target qubit. Consequently, at the output of the quantum circuit, we obtain the state

$$|GHZ\rangle = \frac{1}{\sqrt{2}}(|000\rangle + |111\rangle).$$

60. *Answer.*
Let us visualize the corresponding quantum circuit:

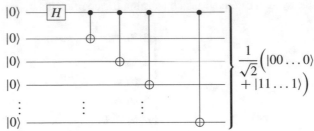

62. *Solution.*
Let us write the expansion $|\psi\rangle$ on a computational basis:

$$|\psi\rangle = a_0 |000\rangle + a_1 |001\rangle + a_2 |010\rangle + a_3 |011\rangle$$
$$+ a_4 |100\rangle + a_5 |101\rangle + a_6 |110\rangle + a_7 |111\rangle.$$

The result of measuring the second qubit, equal to $|0\rangle$, will be obtained with a probability $p = |a_0|^2 + |a_1|^2 + |a_4|^2 + |a_5|^2$. Taking account of $a_0 = 2/5$, $a_1 = 0$, $a_4 = 0$, and $a_5 = 2/5$, we obtain $p = (2/5)^2 + (2/5)^2 = 8/25$.

63. *Answer.* The entangled states are $|\psi_2\rangle$, $|\psi_3\rangle$, $|\psi_4\rangle$.

64. *Answer.* The entangled state is $|\psi_1\rangle$, the state $|\psi_2\rangle$ can be factored:

$$|\psi_2\rangle = \left[\frac{1}{\sqrt{2}}(|0\rangle + |1\rangle)\right]\left[\frac{1}{\sqrt{2}}(-|0\rangle + |1\rangle)\right].$$

and therefore is not entangled.

65. *Solution.*
Zhegalkin polynomial has the following form

$$G(x_1, x_2) = g_0 \wedge 1 \oplus g_1 \wedge x_1 \oplus g_2 \wedge x_2 \oplus g_3 \wedge x_1 \wedge x_2,$$

where $g_0, \ldots, g_3 \in \mathbb{B}$.
Using the conditions $f(0, 0) = 0$, $f(0, 1) = 1$, $f(1, 0) = 0$, $f(1, 1) = 0$, we will obtain a system of equations with respect to unknown coefficients $g_0, \ldots, g_3$:

$$\begin{cases} g_0 = 0, \\ g_0 \oplus g_2 = 1, \\ g_0 \oplus g_1 = 0, \\ g_0 \oplus g_1 \oplus g_2 \oplus g_3 = 0. \end{cases}$$

The solution of this system are Boolean values $g_0 = 0$, $g_1 = 0$, $g_2 = 1$, $g_3 = 1$. Finally, we write down the ANF:

$$f(x_1, x_2) = x_2 \oplus x_1 x_2.$$

66. *Answer.*
(1) $g(x_1, x_2, x_3) = x_2 \oplus (x_1 \wedge x_3)$;
(2) $h(x_1, x_2, x_3, x_4) = 1 \oplus x_1 \oplus x_3 \oplus (x_1 \wedge x_2 \wedge x_3)$.

67. *Answer.*

(1) $g(x_1, x_2, x_3) = 1 \oplus x_1 \oplus x_3 \oplus (x_1 \wedge x_2) \oplus (x_1 \wedge x_3)$;

(2) $h(x_1, x_2, x_3, x_4) = 1 \oplus (x_1 \wedge x_3) \oplus (x_1 \wedge x_4) \oplus (x_1 \wedge x_2 \wedge x_3) \oplus (x_1 \wedge x_2 \wedge x_4)$.

68. *Answer.*

(1) $f(x_1, x_2) = 1 \oplus x_1 \oplus x_2 \oplus x_3 \oplus x_1 x_2 \oplus x_1 x_3 \oplus x_2 x_3$;

(2) $g(x_1, x_2, x_3) = x_1 \oplus x_2 \oplus x_3 \oplus x_1 x_2 \oplus x_1 x_3 \oplus x_2 x_3$.

69. *Answer.*

(1) $f(x_1, x_2) = 1 \oplus x_1 \oplus x_1 x_2$;

(2) $f(x_1, x_2, x_3) = x_1 \oplus x_3 \oplus x_1 x_2 \oplus x_1 x_3 \oplus x_1 x_2 x_3$;

(3) $f(x_1, x_2, x_3, x_4) = 1 \oplus x_1 \oplus x_3 \oplus x_1 x_2 \oplus x_1 x_3 \oplus x_1 x_4 \oplus x_3 x_4 \oplus x_1 x_2 x_3 \oplus x_1 x_2 x_4 \oplus x_1 x_3 x_4 \oplus x_1 x_2 x_3 x_4$;

(4) $f(x_1, x_2, x_3, x_4) = x_1 \oplus x_3 \oplus x_5 \oplus x_1 x_2 \oplus x_1 x_3 \oplus x_1 x_4 \oplus x_1 x_5 \oplus x_3 x_4 \oplus x_3 x_5 \oplus x_1 x_2 x_3 \oplus x_1 x_2 x_4 \oplus x_1 x_2 x_5 \oplus x_1 x_3 x_4 \oplus x_1 x_3 x_5 \oplus x_1 x_4 x_5 \oplus x_3 x_4 x_5 \oplus x_1 x_2 x_3 x_4 \oplus x_1 x_2 x_3 x_5 \oplus x_1 x_2 x_4 x_5 \oplus x_1 x_3 x_4 x_5 \oplus x_1 x_2 x_3 x_4 x_5$.

70. *Answer.* $P(x) = 7 - x$, where $0 \leqslant x \leqslant 7$.

71. *Answer.*

x	0	1	2	3	4	5	6	7
$g(x)$	0	0	0	0	1	1	3	7

72. *Solution.*

Let us make a table of the values of the function $f(x)$ (Table 10.1).

To code the argument, we use three qubits $|x_1\rangle$, $|x_2\rangle$, and $|x_3\rangle$. Since $f(7) = 81$ and $2^6 < 81 < 2^7$, then seven qubits are required to code the response: $|q_1\rangle - |q_7\rangle$. The first bit of the response takes the values (0000 0011) and is therefore implemented using the function $q_1 = x_1 x_2$.

The second bit, (0000 1100), the function is $q_2 = x_1 \oplus x_1 x_2$.

The third bit, (0011 0101), the function is $q_3 = x_1 x_2 \oplus x_1 x_3 \oplus x_2$.

The fourth bit, (0101 0000), the function is $q_4 = x_1 x_3 \oplus x_3$.

The fifth bit, (1000 1000), the function is $q_5 = 1 \oplus x_2 \oplus x_3 \oplus x_2 x_3$.

The sixth bit, (0000 0000), the function is $q_6 = 0$.

The seventh bit, (0101 0101), the function is $q_7 = x_3$.

We obtain the following circuit:

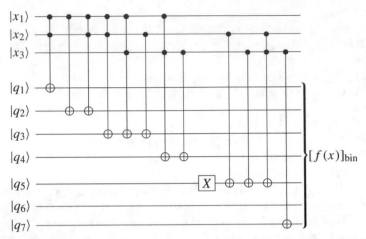

This circuit implements the integer function $f : [0, 7] \rightarrow [4, 81]$, acting by the rule $f(x) = (x + 2)^2$.

73. *Solution.*

For simplicity, we will number the planets of the Solar System from zero: Mercury—zeroth, Venus—first, etc., up to Neptune—seventh planet. This way of numbering allows only three qubits to be used to code the input data.

The data on the number of satellites $s(n), n = 0, 1, \ldots, 7$, known at the moment of this section preparation, are presented in Table 10.2.

As you can see from Table 10.2, the function $s(n)$ reaches its maximal value at $n = 5$: $s(5) = 82$. Consequently, seven qubits are needed to code the response. Let us make a table of values of the function $s(n)$, in which we specify the binary representation of the argument $(n)_{bin}$.

n	$(n)_{bin}$	$s(n)$	$[s(n)]_{bin}$
0	000	0	0000000
1	001	0	0000000
2	010	1	0000001
3	011	2	0000010
4	100	79	1001111
5	101	82	1010010
6	110	27	0011011
7	111	14	0001110
qubits	$x_1x_2x_3$		$q_1q_2q_3q_4q_5q_6q_7$

The first bit of the response $(0000\,1100)$ is implemented using the function $q_1 = x_1 \oplus x_1x_2$.

The second bit, $(0000\,0000)$, function $q_2 = 0$.

The third bit, $(0011\,0101)$, function $q_3 = x_1x_2 \oplus x_1x_3$.

The fourth bit, $(0000\,1011)$, function $q_4 = x_1x_2x_3 \oplus x_1x_3 \oplus x_1$.

Table 10.1 Table of the values of the function

x	$(x)_{\text{bin}}$	$f(x)$	$[f(x)]_{\text{bin}}$
0	000	4	0000100
1	001	9	0001001
2	010	16	0010000
3	011	25	0011001
4	100	36	0100100
5	101	49	0110001
6	110	64	1000000
7	111	81	1010001
qubits	$x_1 x_2 x_3$		$q_1 q_2 q_3 q_4 q_5 q_6 q_7$

Table 10.2 Number of satellites of Solar System planets

n	0	1	2	3	4	5	6	7
$s(n)$	0	0	1	2	79	82	27	14

The fifth bit, (0000 1001), function $q_5 = x_1 x_2 \oplus x_1 x_3 \oplus x_1$.

The sixth bit, (0001 1111), function $q_6 = x_1 x_2 x_3 \oplus x_2 x_3 \oplus x_1$.

The seventh bit, (0010 1010), function $q_7 = x_1 x_2 x_3 \oplus x_1 x_2 \oplus x_1 x_3 \oplus x_2 x_3 \oplus x_1 \oplus x_2$.

We obtain the following circuit:

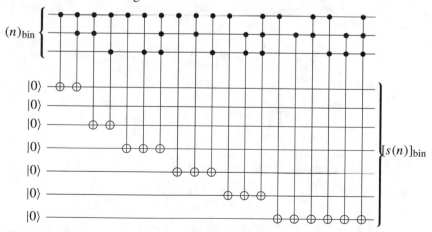

74. *Proof.*

(a) According to the introduced definition,

$$\omega_k = (e^{2\pi i})^{k/N} = e^{2\pi i k/N}, \quad k = 0, 1, \ldots, N-1.$$

The module of the complex number $\omega_k = e^{2\pi i k/N}$ is equal to one for all values of the variable k and the argument is equal to $\arg \omega_k = 2\pi k/N, k = 0, 1, \ldots, N-$

Fig. 10.1 To Exercise 74.
Location of the roots of the
Nth degree on the unit circle
for $N = 9$

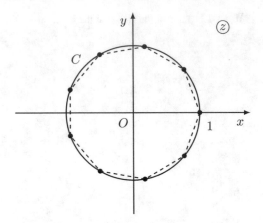

1. It can, therefore, be concluded that the roots of the Nth degree of the one
are on the unit circle C, where the first root is ω_0, corresponding $k = 0$, lies on
the real axis, and ω_k are dividing the circle by N arcs of equal length (see the
example for the special case $N = 9$ in Fig. 10.1).

(b) Convert the representative form of recording the number $\omega_{k+N/2}$:

$$\omega_{k+N/2} = e^{\frac{2\pi i(k+N/2)}{N}} = e^{\frac{2\pi i k}{N}} e^{\pi i} = \omega_k e^{\pi i}.$$

Using equality $e^{\pi i} = -1$, we obtain: $\omega_{k+N/2} = -\omega_k$ for even N and $k = 0, 1, \ldots, N/2 - 1$.

(c) The values ω_k form a geometric progression, the denominator of which is
equal to $\omega_1 = e^{2\pi i/N}$. Using the formula for the sum of the geometric progression, we obtain

$$\sum_{k=0}^{N-1} \omega_k = \sum_{k=0}^{N-1} e^{2\pi i k/N} = \frac{(e^{2\pi i/N})^N - 1}{e^{2\pi i/N} - 1} = 0.$$

(d)

$$\prod_{k=0}^{N-1} \omega_k = \prod_{k=0}^{N-1} e^{2\pi i k/N} = e^{\sum_{k=0}^{N-1} 2\pi i k/N} = e^{(2\pi i \sum_{k=0}^{N-1} k)/N}.$$

The sum in the exponent indicator is equal to $\sum_{k=0}^{N-1} k = \dfrac{N(N-1)}{2}$, which means
that

$$\prod_{k=0}^{N-1} \omega_k = e^{\pi i(N-1)} = \cos \pi(N-1) + i \sin \pi(N-1) = (-1)^{N-1}.$$

75. *Solution.*

(a) Consider a polynomial $f(z) = z^N - 1$. Expansion of $f(z)$ into multipliers of the kind $(z - \omega_k)$, where ω_k are zeros of this polynomial or, in other words, roots of unity, leads to equality:

$$\prod_{k=0}^{N-1}(z - \omega_k) = z^N - 1.$$

(b) Writing down the roots of unity ω_k as $\omega_k = e^{2\pi i k/N}$, rewrite the sum on the left side of the equality as follows:

$$\sum_{k=0}^{N-1}(\omega_k)^d = \sum_{k=0}^{N-1} e^{2\pi i k d/N}.$$

Next, let us consider two cases: $0 < d < N$ and $d = N$.
For the first case we will use the formula for the sum of geometric progression

$$1 + q + q^2 + \cdots + q^{N-1} = \frac{q^N - 1}{q - 1}, \text{ where } q \neq 1.$$

$$\sum_{k=0}^{N-1}(\omega_k)^d = \frac{e^{(2\pi i k d/N)N} - 1}{e^{2\pi i d/N} - 1} = 0.$$

If $d = N$, then all summands in the analyzed sum are equal to one. Finally, we obtain

$$\sum_{k=0}^{N-1}(\omega_k)^d = \begin{cases} 0, & 1 \leqslant d \leqslant N - 1; \\ N, & d = N. \end{cases}$$

76. *Proof.*

Prove, for example, the last of the equations given in the exercise:

$$\sum_{k=0}^{N-2}\sum_{k'=1}^{N-1} \omega^{2k} = \sum_{k=0}^{N-2}(N - k - 1)\omega^{2k}$$

$$= (N - 1)\sum_{k=0}^{N-2} \omega^{2k} - \sum_{k=0}^{N-2} k\omega^{2k};$$

We use an auxiliary identity $\sum_{k=0}^{N} ka^k = a\dfrac{d}{dt}\dfrac{1 - a^{N+1}}{1 - a}$ for the final sum:

$$\sum_{\substack{k,k'=0 \\ k<k'}}^{N-1} \frac{\omega_k\omega_{k'}}{\omega_{k'-k}} = (N - 1)\frac{1 - \omega^{2(n-1)}}{1 - \omega^2}$$

$$- \frac{\omega^2}{(1 - \omega^2)^2}\left(1 + \frac{N - 2}{\omega^2} - \frac{N - 1}{\omega^4}\right) = \frac{N}{1 - \omega^2}.$$

The other points of the exercise are proved in the same way.

77. *Proof.*

Using the definition of the classic discrete Fourier transform (8.4), we convert the value $|y_k|^2$ for all $k = 0, 1, \ldots, N-1$:

$$|y_k|^2 = \left(\sum_{j_1=0}^{N-1} x_{j_1} e^{2\pi i j_1 k/N} \right) \left(\sum_{j_2=0}^{N-1} \overline{x}_{j_2} e^{-2\pi i j_2 k/N} \right)$$

$$= \sum_{j_1, j_2=0}^{N-1} x_{j_1} \overline{x}_{j_2} e^{2\pi i (j_1 - j_2) k/N}.$$

Next, let us calculate the sum on the right side of the formula, which is the content of the Parseval theorem (10.38):

$$\sum_{k=0}^{N-1} |y_k|^2 = \sum_{k=0}^{N-1} \sum_{j_1, j_2=0}^{N-1} x_{j_1} \overline{x}_{j_2} e^{2\pi i (j_1 - j_2) k/N}$$

$$= \sum_{j_1, j_2=0}^{N-1} x_{j_1} \overline{x}_{j_2} \sum_{k=0}^{N-1} e^{2\pi i (j_1 - j_2) k/N}.$$

The internal sum is easily calculated (see formula (8.9)):

$$\sum_{k=0}^{N-1} e^{2\pi i (j_1 - j_2) k/N} = N \delta_{j_1 j_2},$$

where $\delta_{j_1 j_2}$ is the Kronecker symbol, which is defined on page 3. As a result, the sum of squares of modules of y_k is equal to

$$\sum_{k=0}^{N-1} |y_k|^2 = N \sum_{j_1, j_2=0}^{N-1} \delta_{j_1 j_2} x_{j_1} \overline{x}_{j_2} = N \sum_{k=0}^{N-1} |x_k|^2.$$

The Parseval theorem is proved.

78. *Answer.*

(1) $\mathcal{F}[\mathbf{x}] = (2, 0, -2, 0)$;

(2) $\mathcal{F}[\mathbf{x}] = (2, 0, 2, 0)$.

79. *Solution.*

According to the definition of the DFT

$$(\mathcal{F}[\mathbf{x}])_n = \sum_{k=0}^{N-1} e^{\frac{2\pi i}{N} kn} C(N-1, n) \quad \text{for } 0 \leqslant n \leqslant N-1.$$

The resulting ratio for n vector component $\mathcal{F}[\mathbf{x}]$ can be converted using Newton binomial formula:

$$(\mathcal{F}[\mathbf{x}])_n = \sum_{k=0}^{N-1} C(N-1, k) \left(e^{\frac{2\pi i n}{N}} \right)^k = \left(1 + e^{\frac{2\pi i n}{N}} \right)^{N-1} = (1 + \omega^n)^{N-1}.$$

for all $n = 0, 1, \ldots, N-1$.

80. *Answer.*

$$(1)\ (\mathcal{F}[\mathbf{x}])_n = \begin{cases} \dfrac{N}{2}, & \text{if } n = 0; \\[2mm] -\dfrac{1 - (-1)^n}{1 - \omega^n}, & \text{if } n = 1, 2, \ldots, N - 1; \end{cases}$$

$$(2)\ (\mathcal{F}[\mathbf{x}])_n = \begin{cases} \dfrac{N}{2}, & \text{if } n = 0; \\[2mm] \dfrac{1 - (-1)^n}{1 - \omega^n}, & \text{if } n = 1, 2, \ldots, N - 1. \end{cases}$$

82. *Solution.*

The problem can be solved by several methods.

First method

Let us use the definition of a quantum discrete Fourier transform (8.15):

$$|00\ldots0\rangle \rightarrow \mathcal{F}[|00\ldots0\rangle] = \frac{1}{\sqrt{2^n}} \sum_{k=0}^{2^n-1} e^{2\pi i jk/2^n} |k\rangle .$$

The state $|00\ldots0\rangle$ in a computational basis has the sequence number $j = 0$, therefore for all $k = 0, 1, \ldots, 2^n - 1$ is valid the equality $e^{2\pi i jk/2^n} = 1$. Therefore

$$\mathcal{F}[|00\ldots0\rangle] = \frac{1}{\sqrt{2^n}} \sum_{k=0}^{2^n-1} |k\rangle = \frac{1}{\sqrt{2^n}} (|0\rangle + |1\rangle + \cdots + |2^n - 1\rangle).$$

Second method

Using the matrix representation of the quantum discrete Fourier transform (8.17), we obtain

$$|00\ldots0\rangle \rightarrow \mathcal{F}[|00\ldots0\rangle]$$

$$= \frac{1}{2^{n/2}} \begin{bmatrix} 1 & 1 & 1 & \cdots & 1 \\ 1 & \omega_n & \omega_n^2 & \cdots & \omega_n^{N-1} \\ 1 & \omega_n^2 & \omega_n^4 & \cdots & \omega_n^{2(N-1)} \\ 1 & \omega_n^3 & \omega_n^6 & \cdots & \omega_n^{3(N-1)} \\ \cdots\cdots\cdots\cdots\cdots \\ 1 & \omega_n^{N-1} & \omega_n^{2(N-1)} & \cdots & \omega_n^{(N-1)(N-1)} \end{bmatrix} \times \begin{bmatrix} 1 \\ 0 \\ 0 \\ 0 \\ \cdots \\ 0 \end{bmatrix}$$

$$= \frac{1}{2^{n/2}} \begin{bmatrix} 1 \\ 1 \\ 1 \\ 1 \\ \vdots \\ 1 \end{bmatrix}.$$

Such a vector corresponds to the state

$$\mathcal{F}[|00\ldots0\rangle] = \frac{1}{\sqrt{2^n}} (|0\rangle + |1\rangle + \cdots + |2^n - 1\rangle).$$

Third method

For calculations, we will use the formula (8.21). Since $j_1 = j_2 = \cdots = j_n = 0$, all multipliers of the form $e^{2\pi i (0.j_n j_{n-1}\cdots)}$ are converted to one

$$\mathcal{F}[|00\ldots0\rangle] = \frac{1}{\sqrt{2^n}}(|0\rangle + |1\rangle)(|0\rangle + |1\rangle)\ldots(|0\rangle + |1\rangle).$$

By removing parentheses, we obtain the final answer

$$\mathcal{F}[|00\ldots0\rangle] = \frac{1}{\sqrt{2^n}}(|00\ldots1\rangle + |00\ldots1\rangle + \cdots + |11\ldots1\rangle)$$

$$= \frac{1}{\sqrt{2^n}}(|0\rangle + |1\rangle + \cdots + |2^n - 1\rangle).$$

83. *Answer.* $\mathcal{F}[|\psi\rangle] = |0\rangle$.

84. *Answer.* $\mathcal{F}[|\psi\rangle] = \frac{1}{\sqrt{2}}(|1\rangle + |N - 1\rangle)$.

85. *Note.*

Taking as a basis the quantum circuit of the direct Fourier transform in Fig. 8.1, let us reverse the order in time of all elementary transformations (i.e., rewrite the circuit "from right to left") and perform the operation of the Hermitian conjugation over each of the operators: $\forall U\ (U \to U^\dagger)$. As a result we obtain the inverse Fourier Transform circuit $\mathcal{F}^{-1}[y] \equiv \mathcal{F}^\dagger[y]$.

86. *Solution.*

The execution time of the classic Fast Fourier Transform algorithm is determined by an asymptotic ratio $T_{\text{cl.}} = \tau_{\text{cl.}}\Theta(N \log_2 N)$, where N is the size of the input vector, and $\tau_{\text{cl.}}$ is the duration of one elementary operation of the algorithm. The Quantum Fourier Transform algorithm is based on a circuit that contains $\Theta\left((\log_2 N)^2\right)$ elementary gates. Assuming that the execution time of a single gate is a constant value $\tau_{\text{quant.}} = \Theta(1)$, we obtain a quantum algorithm speedup $S = \dfrac{T_{\text{cl.}}}{T_{\text{quant.}}}$ comparing to the classic case:

$$S = \frac{\tau_{\text{cl.}}\Theta(N \log_2 N)}{\tau_{\text{cl.}}\Theta\left((\log_2 N)^2\right)} = \Theta\left(\frac{N}{\log_2 N}\right).$$

87. *Solution.*

The asymptotic complexity of the parallel version of the Fast Fourier Transform algorithm presented on page 47, is equal to $\Theta(\sqrt{N} \log_2 N)$. By performing calculations similar to the solution in the previous exercise, we obtain the speedup S achieved by the quantum algorithm of the Fourier transform compared to the parallel algorithm of the Fast Fourier transform:

$$S = \frac{\Theta(\sqrt{N} \log_2 N)}{\Theta\left((\log_2 N)^2\right)} = \Theta\left(\frac{\sqrt{N}}{\log_2 N}\right).$$

References

1. Greenberger DM, Horne M, Zeilinger A (1989) Going beyond Bell's theorem. In Kafatos M (ed) Bell's theorem, quantum theory and conceptions of the universe. Fundamental theories of physics, vol 37. Springer, Dordrecht, pp 69–72
2. Gamelin TW (2001) Complex analysis (Series: Undergraduate texts in mathematics). Springer, 478 p
3. Jozsa R (1994) Fidelity for mixed quantum states. J Modern Opt 41(12):2315–2323
4. Liang YC, Yeh Yu-H, Mendonça PEMF, Teh RY, Reid MD, Drummond PD (2019) Quantum fidelity measures for mixed states. Rep Prog Phys 82(2):07600

Appendix A: The Postulates of Quantum Theory

A

Quantum theory is a theory that establishes the way of describing laws of motion of physical systems, for which the values characterizing the system and having the dimension of the action are comparable to the Planck[1] constant, or, as they say, the **quantum of action**

$$h = 6.626\ldots \times 10^{-34} \text{ J} \cdot \text{sec.} \tag{A.1}$$

In particular, quantum theory describes the processes taking place in *microworld*, to which objects belong *structural elements of matter*:

- atoms,
- molecules,
- primitive crystal cells,
- atomic nuclei, and
- elementary particles.

These elements of the microworld are characterized by very peculiar laws of motion. These laws differ significantly from the laws of classical mechanics, describing mechanical motion in classical physics. A number of effects (superconductivity, superfluidity, ferromagnetism) as well as physicochemical properties of matter can only be quantitatively explained within the framework of quantum theory.

Let us present the main provisions of quantum theory as a set of postulates (see Fig. A.1).

Postulate about the state. The state of the quantum system is fully specified by the state vector, which has the properties of uniqueness (with the accuracy to an arbitrary phase) and normality.

Postulate about operators of physical quantities. Each physical quantity corresponds to a linear Hermitian operator, in which eigenvalues are possible values of the physical quantity, and its eigenvectors are states corresponding to the indicated eigenvalue.

---

[1]Max Karl Ernst Ludwig Planck (1858–1947).

© The Editor(s) (if applicable) and The Author(s), under exclusive license to Springer Nature Switzerland AG 2021
S. Kurgalin and S. Borzunov, *Concise Guide to Quantum Computing*,
Texts in Computer Science, https://doi.org/10.1007/978-3-030-65052-0

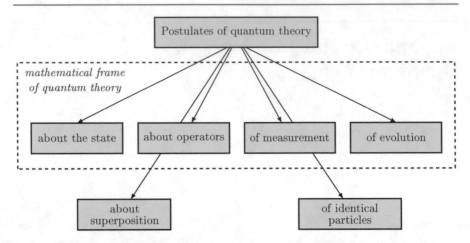

Fig. A.1 Quantum theory postulates system

Postulate of measurement. In the decomposition of an arbitrary vector $|\psi\rangle$ by the orthonormalized system of eigenvectors $|\psi_k\rangle$ of physical quantity F:

$$|\psi\rangle = \sum_k a_k \, |\psi_k\rangle \tag{A.2}$$

the values $|a_k|^2$ are equal to the probabilities of finding the system in states $|\psi_k\rangle$, i.e., to the probabilities that when measuring F its value will be the kth eigenvalue.

The postulate of measurement is also referred as the **Born**[2] **rule**.

Postulate of evolution. The evolution of the state vector is described by the Schrödinger[3] time-dependent equation:

$$i\hbar \frac{\partial}{\partial t} \, |\psi\rangle = \mathcal{H} \, |\psi\rangle \,, \tag{A.3}$$

where $\mathcal{H}$ is the Hamiltonian[4] operator of quantum system, $\hbar = h/2\pi$ is a fundamental physical constant, which is called, as well as (A.1), the Planck constant.

The above postulates determine the mathematical apparatus of quantum theory. Two other statements are used to fill it with physical content: the principle of superposition of states and the postulate of identical particles (see Fig. A.1) [1–3].

Superposition of state postulate. If a quantum system can be in the states $|\psi_1\rangle$, $|\psi_2\rangle$, ..., $|\psi_n\rangle$, in which the physical value F takes values $f_1, f_2, ..., f_n$, respectively, the system can be in the state $|\Psi\rangle$, which is a linear superposition of states $|\psi_k\rangle$:

$$|\Psi\rangle = \sum_{k=1}^n a_k \, |\psi_k\rangle \,. \tag{A.4}$$

---

[2]Max Born (1882–1970).
[3]Erwin Rudolf Josef Alexander Schrödinger (1887–1961).
[4]William Rowan Hamilton (1805–1865).

Postulate of identical particles. Particles with integer spin are described by symmetrical wave functions (are bosons[5]), particles with a half-integer spin are described by symmetric wave functions (are fermions[6]).

Apparently, all particles occurring in nature belong either to the class of fermions or to the class of bosons.

---

[5] Satyendra Nath Bose (1894–1974).
[6] Enrico Fermi (1901–1954).

Appendix B: Complex Matrices

In quantum theory and its applications among matrices with complex coefficients, the classes of Hermitian and unitary matrices play a special role. Hermitian matrices correspond to the operators of observed physical quantities. In its turn, unitary matrices correspond to linear transformations preserving vector lengths and, consequently, the normalization of wave functions [2].

B.1 Hermitian Matrices

Consider a matrix Z of size $m \times n$, containing complex elements $Z = (z_{ij})$, where $i = 1, 2, \ldots, m$, $j = 1, 2, \ldots, n$. **Hermitian conjugate** matrix with reference to Z is called matrix $Z^\dagger$, which elements are equal to

$$z_{ij}^\dagger = z_{ji}^*. \tag{B.1}$$

In order to obtain a Hermitian conjugate matrix, transpose and complex conjugation operations are applied to Z. These operations are independent and can be performed in any order.

Example B.1 Hermitian conjugate matrix of $Z = \begin{bmatrix} 1+i & 2+3i \\ -1 & 5-4i \end{bmatrix}$ is the matrix

$$Z^\dagger = \left(\begin{bmatrix} 1+i & 2+3i \\ -1 & 5-4i \end{bmatrix}^T \right)^* = \left(\begin{bmatrix} 1+i & -1 \\ 2+3i & 5-4i \end{bmatrix} \right)^* = \begin{bmatrix} 1-i & -1 \\ 2-3i & 5+4i \end{bmatrix}. \tag{B.2}$$

$\square$

Note. Sometimes for the notation of the Hermitian conjugate matrix is used Z^+ or Z^H.

© The Editor(s) (if applicable) and The Author(s), under exclusive license to Springer Nature Switzerland AG 2021
S. Kurgalin and S. Borzunov, *Concise Guide to Quantum Computing*,
Texts in Computer Science,

Let us list the main properties of a Hermitian conjugate operation:

(1) $I^\dagger = I$;

(2) $(Z_1 + Z_2)^\dagger = Z_1^\dagger + Z_2^\dagger$;

(3) $(\lambda Z)^\dagger = \lambda^* Z^\dagger \quad \forall \lambda \in \mathbb{C}$;

(4) $(Z^\dagger)^\dagger = Z$;

(5) if there exists A^{-1}, then $(A^{-1})^\dagger = (A^\dagger)^{-1}$;

(6) $\det A^\dagger = \det A^* = (\det A)^*$.

Theorem B.1 *If for complex matrices Z_1 and Z_2 the product $Z_1 Z_2$ is defined, then*

$$(Z_1 Z_2)^\dagger = Z_2^\dagger Z_1^\dagger. \tag{B.3}$$

Proof.

The validity of the theorem follows from the property of transposition of the matrix product:

$$(Z_1 Z_2)^T = Z_2^T Z_1^T. \tag{B.4}$$

With the help of Eq. (B.4), we obtain a chain of equations

$$(Z_1 Z_2)^\dagger = ((Z_1 Z_2)^T)^* = (Z_2^T Z_1^T)^* = (Z_2^T)^* (Z_1^T)^* = Z_2^\dagger Z_1^\dagger, \tag{B.5}$$

which proves the identity (B.3).

Among complex matrices, Hermitian matrices are very widely used. **Hermitian** matrix is a square matrix with $Z^\dagger = Z$. The corresponding condition for the elements of such a matrix is $\forall i, j \ (z_{ij} = z_{ji}^*)$. In other words, the Hermitian matrix coincides with its Hermitian conjugate one [4].

Example B.2 Matrix $Z = \begin{bmatrix} 1 & -3-i \\ -3+i & 1 \end{bmatrix}$ is as easy to see, Hermitian. Let us check it out.

$$Z^\dagger = \left(\begin{bmatrix} 1 & -3-i \\ -3+i & 1 \end{bmatrix}^T \right)^*$$

$$= \left(\begin{bmatrix} 1 & -3+i \\ -3-i & 1 \end{bmatrix} \right)^* = \begin{bmatrix} 1 & -3-i \\ -3+i & 1 \end{bmatrix}. \tag{B.6}$$

It should be reminded that the application of the complex conjugation operation to the real number leaves this number unchanged.   □

Example B.3 Gell-Mann[7] matrices λ_k, where $k = 1, 2, \ldots, 8$,

$$\lambda_1 = \begin{bmatrix} 0 & 1 & 0 \\ 1 & 0 & 0 \\ 0 & 0 & 0 \end{bmatrix}, \quad \lambda_2 = \begin{bmatrix} 0 & -i & 0 \\ i & 0 & 0 \\ 0 & 0 & 0 \end{bmatrix}, \quad \lambda_3 = \begin{bmatrix} 1 & 0 & 0 \\ 0 & -1 & 0 \\ 0 & 0 & 0 \end{bmatrix},$$

$$\lambda_4 = \begin{bmatrix} 0 & 0 & 1 \\ 0 & 0 & 0 \\ 1 & 0 & 0 \end{bmatrix}, \quad \lambda_5 = \begin{bmatrix} 0 & 0 & -i \\ 0 & 0 & 0 \\ i & 0 & 0 \end{bmatrix}, \quad \lambda_6 = \begin{bmatrix} 0 & 0 & 0 \\ 0 & 0 & 1 \\ 0 & 1 & 0 \end{bmatrix},$$

$$\lambda_7 = \begin{bmatrix} 0 & 0 & 0 \\ 0 & 0 & -i \\ 0 & i & 0 \end{bmatrix}, \quad \lambda_8 = \frac{1}{\sqrt{3}} \begin{bmatrix} 1 & 0 & 0 \\ 0 & 1 & 0 \\ 0 & 0 & -2 \end{bmatrix}, \tag{B.7}$$

are for all k, as it is easy to notice, Hermitian. Gell-Mann matrices are a generalization of Pauli matrices to describe the quark model of strongly interacting particles.

The set $\Lambda = \{I, \lambda_1, \lambda_2, \ldots, \lambda_8\}$ forms the basis in the vector space of complex matrices of size 3×3. In fact, an arbitrary matrix

$$M = \begin{bmatrix} m_{11} & m_{12} & m_{13} \\ m_{21} & m_{22} & m_{23} \\ m_{31} & m_{32} & m_{33} \end{bmatrix}$$

can be represented as a linear combination of

$$M = a_0 I + \sum_{i=1}^{8} a_i \lambda_i, \tag{B.8}$$

where $a_{0-8} \in \mathbb{C}$ are some complex numbers. Using a method similar to the one presented in the solution of the Exercise 10, we obtain a vector of expansion coefficients:

$$(a_0, a_1, \ldots, a_8)$$
$$= \Big(\frac{1}{3}(m_{11} + m_{22} + m_{33}), \frac{1}{2}(m_{12} + m_{21}), \frac{i}{2}(m_{12} - m_{21}),$$
$$\frac{1}{2}(m_{11} - m_{22}), \frac{1}{2}(m_{13} + m_{31}), \frac{i}{2}(m_{13} - m_{31}),$$
$$\frac{1}{2}(m_{23} + m_{32}), \frac{i}{2}(m_{23} - m_{32}), \frac{\sqrt{3}}{6}(m_{11} + m_{22} - 2m_{33}) \Big).$$

The set $\Lambda = \{I, \lambda_1, \lambda_2, \ldots, \lambda_8\}$ is called **Gell-Mann basis** of vector space of complex matrices of the second order.

$\square$

Example B.4 Show that if Z_1 and Z_2 are complex matrices of the same order, the matrix $\frac{1}{2}(Z_1 Z_2 + Z_2 Z_1)$ is Hermitian.

---

[7]Murray Gell-Mann (1929–2019).

Proof.

Let us introduce the notation $W = \frac{1}{2}(Z_1 Z_2 + Z_2 Z_1)$ and find Hermitian conjugate matrix with reference to W:

$$W^\dagger = \left(\frac{1}{2}(Z_1 Z_2 + Z_2 Z_1)\right)^\dagger = \frac{1}{2}((Z_1 Z_2)^\dagger + (Z_2 Z_1)^\dagger)$$

$$= \frac{1}{2}(Z_2 Z_1 + Z_1 Z_2) = W. \tag{B.9}$$

Thus, it is proved that W is Hermitian matrix.   □

Note. Hermitian matrices are also called **self-conjugated** matrices. The theory of self-conjugated matrices is widely used in modern physics [2].

B.2 Unitary Matrices

A square matrix U with complex elements is called **unitary**, if the condition $U^\dagger U = I$ is fulfilled. The unitarity condition can be written in other equivalent forms as follows:

$$UU^\dagger = I \quad \text{or} \quad U^\dagger = U^{-1}. \tag{B.10}$$

Example B.5 Let us prove that matrix $Z = \frac{1}{\sqrt{2}}\begin{bmatrix} 1 & 1 \\ -i & i \end{bmatrix}$ is a unitary one. To do this, let us calculate the product $Z^\dagger Z$:

$$Z^\dagger Z = \frac{1}{\sqrt{2}}\left(\begin{bmatrix} 1 & 1 \\ -i & i \end{bmatrix}\right)^\dagger \times \frac{1}{\sqrt{2}}\begin{bmatrix} 1 & 1 \\ -i & i \end{bmatrix}$$

$$= \frac{1}{\sqrt{2}}\begin{bmatrix} 1 & i \\ 1 & -i \end{bmatrix} \times \frac{1}{\sqrt{2}}\begin{bmatrix} 1 & 1 \\ -i & i \end{bmatrix} = \begin{bmatrix} 1 & 0 \\ 0 & 1 \end{bmatrix}. \tag{B.11}$$

Therefore, Z is unitary matrix.   □

Theorem B.2 *The unitary matrix determinant is a complex number, the modulus of which is equal to one.*

Proof.

Let U is an arbitrary unitary matrix. Using the property (6) of Hermitian conjugation on page 110, we present the modulus of determinant U as follows:

$$|\det U| = \sqrt{(\det U)(\det U)^*} = \sqrt{(\det U)(\det U^\dagger)}. \tag{B.12}$$

Since the product of matrix determinants is equal to the determinant of their product, $\forall A, B \ (\det A \times \det B = \det(AB))$, then

$$|\det U| = \sqrt{\det(UU^\dagger)} = \sqrt{\det I} = 1. \tag{B.13}$$

Thus, the unitary matrix determinant modulus is equal to one. The theorem B.2 about unitary matrix determinant is proved.

There is a close connection between unitary and Hermitian matrices: each unitary matrix A is represented as $A = \exp(iB)$, where B is a Hermitian matrix [5].

Example B.6 What condition must the complex numbers u_1 and u_2 and the real number φ satisfy

$$\begin{bmatrix} u_1 & u_2 \\ -e^{i\varphi}u_2^* & e^{i\varphi}u_1^* \end{bmatrix} \tag{B.14}$$

for the matrix to be unitary?

Solution.

Let us check the unitarity condition $UU^{\dagger} = I$ by direct calculation of the product of the source matrix and the Hermitian conjugated to it.

$$\begin{bmatrix} u_1 & u_2 \\ -e^{i\varphi}u_2^* & e^{i\varphi}u_1^* \end{bmatrix} \times \begin{bmatrix} u_1 & u_2 \\ -e^{i\varphi}u_2^* & e^{i\varphi}u_1^* \end{bmatrix}^{\dagger}$$

$$= \begin{bmatrix} u_1 & u_2 \\ -e^{i\varphi}u_2^* & e^{i\varphi}u_1^* \end{bmatrix} \times \begin{bmatrix} u_1^* & -e^{-i\varphi}u_2 \\ u_2^* & e^{-i\varphi}u_1 \end{bmatrix}$$

$$= \begin{bmatrix} |u_1|^2 + |u_2|^2 & 0 \\ 0 & |u_1|^2 + |u_2|^2 \end{bmatrix}. \tag{B.15}$$

It follows that in order to fulfill the unitarity property, a necessary and sufficient condition is the fulfillment of the equality $|u_1|^2 + |u_2|^2 = 1$. In this connection, the parameter $\varphi \in \mathbb{R}$ can take arbitrary values. $\qquad\square$

Theorem B.3 *Eigenvalues of the Hermitian operator are real.*

Proof.

Let A is an arbitrary Hermitian matrix of size $n \times n$, with its eigenvector b, which corresponds to its eigenvalue λ_0. This means that the equality is fulfilled

$$Ab = \lambda_0 b. \tag{B.16}$$

Consider the expression $b^{\dagger}Ab$. It is equal to the real number, i.e., according to the theorem of the Hermitian conjugation of the product on the page 110

$$(b^{\dagger}Ab)^{\dagger} = b^{\dagger}A^{\dagger}(b^{\dagger})^{\dagger} = b^{\dagger}Ab. \tag{B.17}$$

According to the (B.16), the following equality is fulfilled

$$b^{\dagger}Ab = \lambda_0 b^{\dagger}b = \lambda_0(|b_1|^2 + |b_2|^2 + \cdots + |b_n|^2), \tag{B.18}$$

where $b_1, b_2, \ldots, b_n$ are components of the vector b.

Comparing (B.17) and (B.18), we obtain that $\lambda_0 \in \mathbb{R}$.

Theorem B.3 about eigenvalues of the Hermitian operator is proved.

Theorem B.4 *All eigenvalues of the unitary matrix lie in the complex plane on a unit circle with the center at the origin of the coordinate system.*

Proof.

Let U is an arbitrary unitary matrix with an eigenvector $\boldsymbol{b}$, which corresponds to its eigenvalue λ_0. This means that the following equality is fulfilled:

$$U\boldsymbol{b} = \lambda_0 \boldsymbol{b}. \tag{B.19}$$

Consider the expression $(U\boldsymbol{b})^\dagger (U\boldsymbol{b})$. By virtue of the fact that (B.19), this expression can be represented as follows:

$$(U\boldsymbol{b})^\dagger (U\boldsymbol{b}) = (\lambda_0 \boldsymbol{b})^\dagger (\lambda_0 \boldsymbol{b}) = (\lambda_0^* \boldsymbol{b}^\dagger)(\lambda_0 \boldsymbol{b}) = \lambda_0^* \lambda_0 \boldsymbol{b}^\dagger \boldsymbol{b} = |\lambda_0|^2 \boldsymbol{b}^\dagger \boldsymbol{b}. \tag{B.20}$$

It should be noted that the property (3) of the Hermitian conjugation operation was used in the conversions (see page 110).

On the other hand, based on the theorem about the Hermitian conjugation of the product on page 110 and on the unitarity property $U^\dagger U = I$, we obtain

$$(U\boldsymbol{b})^\dagger (U\boldsymbol{b}) = (\boldsymbol{b}^\dagger U^\dagger)(U\boldsymbol{b}) = \boldsymbol{b}^\dagger (U^\dagger U)\boldsymbol{b} = \boldsymbol{b}^\dagger I \boldsymbol{b} = \boldsymbol{b}^\dagger \boldsymbol{b}. \tag{B.21}$$

Comparing (B.20) and (B.21), we come to the conclusion: $|\lambda_0|^2 = 1$.

Therefore, the complex number λ_0 is located on the complex plane at a distance $\rho = 1$ from the origin of coordinates. The locus of all such points λ_0 is a circle of unit radius with the center at the origin of coordinates, which was to be proved.

Theorem B.4 about eigenvalues of the unitary operator is proved.

Example B.7 Let us calculate eigenvalues and eigenvectors of Pauli matrices σ_1, σ_2, σ_3 (see the definition on the page 7).

Solution.

Let us determine the eigenvalues of the matrix $\sigma_1 = \begin{bmatrix} 0 & 1 \\ 1 & 0 \end{bmatrix}$. For this purpose, we solve the characteristic equation

$$\begin{vmatrix} -\lambda & 1 \\ 1 & -\lambda \end{vmatrix} = 0. \tag{B.22}$$

Expanding the determinant of the second order, we obtain the equation $(-\lambda)^2 - 1 = 0$. The equation with respect to the variable λ has two roots $\lambda_1 = 1$ and $\lambda_2 = -1$, and these numbers are eigenvalues of Pauli matrix σ_1.

Each of its eigenvalues λ_i, where $i = 1, 2$, corresponds to eigenvector $\boldsymbol{b}_i$. First let us calculate $\boldsymbol{b}_1$ for $\lambda_1 = 1$. Denote the unknown complex components of this vector: $\boldsymbol{b}_1 = \begin{bmatrix} z_1 \\ z_2 \end{bmatrix}$. According to the definition of the matrix eigenvector, z_1 and z_2 values satisfy the system of linear equations

$$\begin{bmatrix} -1 & 1 \\ 1 & -1 \end{bmatrix} \begin{bmatrix} z_1 \\ z_2 \end{bmatrix} = 0. \tag{B.23}$$

The resulting system contains two equations

$$\begin{cases} -z_1 + z_2 = 0, \\ z_1 - z_2 = 0. \end{cases} \tag{B.24}$$

which differ only in sign. Consequently, the condition $z_1 = z_2$ is fulfilled, and the vector $\boldsymbol{b}_1$ can be represented as follows:

$$\boldsymbol{b}_1 = \begin{bmatrix} z_1 \\ z_1 \end{bmatrix} = z_1 \begin{bmatrix} 1 \\ 1 \end{bmatrix}, \tag{B.25}$$

where $z_1 \in \mathbb{C}$. Quantum mechanics usually uses normalized vectors, so let us impose an additional condition of normalization:

$$|z_1|^2 + |z_1|^2 = 1. \tag{B.26}$$

For simplicity, let us select one of the real roots of Eq. (B.26) $z_1 = 1/\sqrt{2}$, which results in $\boldsymbol{b}_1 = \dfrac{1}{\sqrt{2}} \begin{bmatrix} 1 \\ 1 \end{bmatrix}$.

Next, the eigenvector which corresponds to $\lambda_2 = -1$ is equal to $\boldsymbol{b}_2 = \dfrac{1}{\sqrt{2}} \begin{bmatrix} 1 \\ -1 \end{bmatrix}$, as it can be easily shown by similar calculations.

Let us formulate a final answer:

for σ_1: $\lambda_{1,2} = \pm 1$, eigenvectors are $\boldsymbol{b}_{1,2} = \dfrac{1}{\sqrt{2}} \begin{bmatrix} 1 \\ \pm 1 \end{bmatrix}$;

for σ_2: $\lambda_{1,2} = \pm 1$, eigenvectors are $\boldsymbol{b}_{1,2} = \dfrac{1}{\sqrt{2}} \begin{bmatrix} 1 \\ \pm i \end{bmatrix}$;

for σ_3: $\lambda_{1,2} = \pm 1$, eigenvectors are $\boldsymbol{b}_1 = \begin{bmatrix} 1 \\ 0 \end{bmatrix}$ and $\boldsymbol{b}_2 = \begin{bmatrix} 0 \\ 1 \end{bmatrix}$. $\qquad \square$

References

1. Dirac PAM (1967) The principles of quantum mechanics, 4th edn. (Series: The international series of monographs on physics, 27). Oxford University Press, 314 p
2. Landau LD, Lifshitz EM (1989) Quantum mechanics: non-relativistic theory. Course of theoretical physics, vol 3, 3rd edn. Pergamon Press, Oxford
3. von Neumann J (2018) Mathematical foundations of quantum mechanics. Princeton University Press, 304 p
4. Kurgalin S, Borzunov S (2021) Algebra and geometry with Python. Springer, Cham. xvi, 425 p
5. Gantmacher FR (2001) The theory of matrices, vol 1. American Mathematical Society, Providence, Rhode Island, x, 374 p

Name Index

Index

© The Editor(s) (if applicable) and The Author(s), under exclusive license to Springer
Nature Switzerland AG 2021
S. Kurgalin and S. Borzunov, *Concise Guide to Quantum Computing*,
Texts in Computer Science, https://doi.org/10.1007/978-3-030-65052-0

Printed in the United States
by Baker & Taylor Publisher Services